WIRELESS LANS DEMYSTIFIED

OTHER BOOKS IN THE DEMYSTIFIED SERIES

3G Wireless Demystified
802.11 Demystified
Bluetooth Demystified
CEBus Demystified
Computer Telephony Demystified
DVD Demystified
GPRS Demystified
MPEG-4 Demystified
SIP Demystified
SONET/SDH Demystified
Streaming Media Demystified
Video Compression Demystified
Videoconferencing Demystified
Wireless Data Demystified
Wireless Messaging Demystified

Wireless LANs Demystified

Jaidev Bhola

McGraw-Hill
New York   Chicago   San Francisco   Lisbon
London   Madrid   Mexico City   Milan   New Delhi
San Juan   Seoul   Singapore   Sydney   Toronto

Cataloging-in-Publication Data is on file with the Library of Congress

McGraw-Hill
A Division of The McGraw·Hill Companies

Copyright © 2002 by The McGraw-Hill Companies, Inc. All rights reserved. Printed in the United States of America. Except as permitted under the United States Copyright Act of 1976, no part of this publication may be reproduced or distributed in any form or by any means, or stored in a data base or retrieval system, without the prior written permission of the publisher.

1 2 3 4 5 6 7 8 9 0   DOC/DOC   0 8 7 6 5 4 3 2

ISBN 0-07-138784-6

The sponsoring editor for this book was Judy Bass and the production supervisor was Pamela A. Pelton. It was set in Century Schoolbook by MacAllister Publishing Services, LLC.

Printed and bound by RR Donnelley.

McGraw-Hill books are available at special quantity discounts to use as premiums and sales promotions, or for use in corporate training programs. For more information, please write to the Director of Special Sales, McGraw-Hill Professional, Two Penn Plaza, New York, NY 10121-2298. Or contact your local bookstore.

> Information contained in this work has been obtained by The McGraw-Hill Companies, Inc. ("McGraw-Hill") from sources believed to be reliable. However, neither McGraw-Hill nor its authors guarantees the accuracy or completeness of any information published herein and neither McGraw-Hill nor its authors shall be responsible for any errors, omissions, or damages arising out of use of this information. This work is published with the understanding that McGraw-Hill and its authors are supplying information but are not attempting to render engineering or other professional services. If such services are required, the assistance of an appropriate professional should be sought.

 This book is printed on recycled, acid-free paper containing a minimum of 50 percent recycled, de-inked fiber.

To Ramya, Sukhwant, and Mom

CONTENTS

Introduction		xiii
Chapter 1	The Wired Enterprise Background	1
	Ethernet	3
	What's in a Name?	5
	Origins of Ethernet	5
	Traditional Wired Ethernet	7
	Fast Wired Ethernet	9
	Gigabit Ethernet	9
	Different Physical Ethernet Network Types	10
	Present-Day Ethernet Networks	11
	802.3	14
	Ethernet in Use	17
	WLAN Advantages	18
	Technical Advantages of WLANs	20
	Summary	22
Chapter 2	Welcome to the Wireless Enterprise	23
	Returning to AlohaNet	26
	Today's WLAN Technology	30
	The 802.11 Standard	30
	802.11b	31
	802.11a	32
	PAN Technology	33
	Bluetooth	33
	Building-to-Building WWANs	34
	In-Building WLANs	36
	Cellular-based Wireless Technology	37
	Wireless Application Protocol (WAP)	37
	CDMA	38
	3G	39
	Summary	44
Chapter 3	Wireless LAN Primer	45
	Physical Media Layer	46
	Radio Frequency (RF)	47

	IR Technology	52
	Microwave	54
	Summary	54
MAC Protocol		55
Wireless Network Infrastructures		57
	Ad Hoc Networking	58
	Infrastructure Networking	62
Selecting the Hardware		63
Choosing the Best Configuration for the Enterprise		64
Coverage Options		64
	Minimal Overlap Coverage Option	64
	Multiple Overlapping Networks Coverage Option	65
	Heavy Overlap Coverage Option	66
Site Surveys		66
Range and Coverage in the Real World		68
Microcells and Roaming		68
Mobile IP		70
	How It Works	71
Security		73
First-generation WLAN Security		74
	SSID	74
	WEP	75
	Authentication	76
A Complete Security Solution		76
Integration with Existing Networks		79
Number of Clients per Access Point		80
Throughput		80
	Getting Killer Throughput	81
Integrity and Reliability of the Wireless Enterprise		83
Summary		83

Chapter 4 Access Point Installation — 85

Access Point Installation and Configuration Outline		87
	Connecting and Powering Up	87
	Assign an IP Address to the Access Point	88
	Using a DHCP Server	88
	Using the Access Point Setup Utility	89
	Setting the Access Point's IP Address and SSID	89
	Verifying the DHCP-Assigned IP Address	89
Access Point Mounting		90

Contents

Key Access Point Features	90
Antennas	90
Ethernet and Serial Ports	91
Infrastructure Network Configuration Examples	91
Root Unit on a Wired LAN	92
Installation Guidelines	94
Basic Guidelines	94
Access Point Utility Software	97
Using the Management Interfaces	98
Configuration	98
Entering Basic Settings	98
Filter Setup	101
Radio Configuration	103
Ethernet Configuration	108
Routing Setup	110
Security Setup	111
Security Overview	111
Setting Up WEP	116
Enabling Additional WEP Security Features	117
Managing Firmware and Configurations	118
Basic Troubleshooting	119
Checking Basic Settings	119
Summary	120

Chapter 5 Wireless Networks and Windows XP 121

Easing the Security Burden with RADIUS	124
Zero Configuration Required	127
Windows XP and Wireless Security	127
Installation and Configuration Guide	128
Wireless Adapters	128
Client Adapter and Possible Network Configurations	131
Preparing for Installation	132
Unpacking the Client Adapter	132
System Requirements	132
Site Requirements	133
Installation of Client Adapters	134
Installing the Driver	134
Installing the Adapter Software Utility	138
Verifying Installation	139
Configuring the Client Adapter	139

Setting RF Network Parameters	139
Setting Advanced Infrastructure Parameters	141
Setting Advanced Ad Hoc Parameters	143
Setting Network Security Parameters	143
Overview of Security Features	145
Using Static WEP	147
Overwriting an Existing Static WEP Key	149
Disabling Static WEP	150
Performing Diagnostics	150
Viewing the Current Status of Your Client Adapter	150
Viewing Statistics for Your Client Adapter	150
Viewing the Link Status Meter	151
Running an RF Link Test	152
Routine Procedures	152
Inserting and Removing a Client Adapter	152
Driver Procedures	153
Troubleshooting	155
Client Adapter Recognition Problems	155
Resolving Resource Conflicts	156
Problems Associating with an Access Point	158
Problems Authenticating with an Access Point	159
Problems Connecting to the Network	159
Configuring the Client Adapter Through Windows XP	159
Overview of Security Features	160
Static WEP Keys	160
EAP (with Static or Dynamic WEP Keys)	160
Configuring the Client Adapter	161
Using Windows XP to Associate with an Access Point	162
Viewing the Current Status of Your Client Adapter	162
Installation and Configuration Guide for Windows CE	163
Hardware Components	163
Software Components	164
Network Configurations Using the Client Adapter	165
Preparing for Installation	167
Unpacking the Client Adapter	167
System Requirements	167
Installing the Client Adapter	168
Installing the Driver and Client Utilities	168
Verifying Installation	172
Enabling Security Features	173
Using Static WEP	173

Contents

Entering a New Static WEP Key and Enabling Static WEP	173
Overwriting an Existing Static WEP Key	176
Disabling Static WEP	176
Windows CE-based Device Notes	176
Using WEP with a Windows CE-based Device	177
Using DHCP with a Windows CE-based Device	177
Advanced Configuration	178
Configuring Your Client Adapter	178
Performing Diagnostics	178
Overview of the Diagnostic Utilities	178
Viewing the Current Status of Your Client Adapter	179
Viewing Statistics for Your Client Adapter	180
Routine Procedures	184
Inserting and Removing a PC Card	184
Upgrading the Client Adapter Software	185
Troubleshooting the Client Adapter	185
Problems Obtaining an IP Address	185
Problems Connecting to the Network	185
Error Messages	186
Getting Help	186
On HPC Devices	186
On PPC Devices	186
Summary	187

Chapter 6 Wireless Home Network Configurations 189

Deployment Considerations	190
Bringing the Wireless Enterprise Home	190
Various WLANs for the Home	190
Ad Hoc Mode Wireless Home LAN Using Windows ICS	191
Infrastructure Mode Wireless Home LAN Using Windows ICS	191
Infrastructure Mode Wireless Home LAN Using a Hardware-Based NAT	192
Choosing Between Ad Hoc and Infrastructure Networks	195
Ad Hoc Network Using ICS	195
Steps to Set Up the NAT Internet Connection and Ad Hoc Wireless Home LAN	196
Test and Troubleshoot the Internet Connection on the Ad Hoc Wireless Home LAN	200
Infrastructure Network Setup Using an Access Point Hub and a Windows ICS NAT	202

Setup of the Buffalo Technologies Airstation WLAR-L11	203
Troubleshooting Issues with the Airstation WLA-L11/ Windows ICS Setup	208
Infrastructure Network Setup Using an Access Point Router	209
Setup of the Buffalo Technologies Airstation WLAR-L11-L	209
Troubleshooting Issues with the Airstation WLAR-L11-L Setup	214
Windows ICS Setup	216
ICS Setup for Windows 2000 and Windows XP	216
Integrating a Wireless Work Network with a Wireless Home Network	218
Overview and Comparison of Wireless Products	220
Windows CE-Based Device Notes	220
Using WEP with a Windows CE-Based Device	220
Connecting to the Corporate Network Through the Home LAN	224
Building the Wireless VPN	225
Summary	225

Chapter 7 Wrap Up   227

Acronyms   231

Glossary   233

Bibliography   243

Index   245

INTRODUCTION

At first glance, creating a wireless enterprise appears to be a straightforward process, and it is. A successful implementation consists in part of planning for future needs and considering industry standards. Even though industry standards continue to evolve, an enterprise built on existing and supported standards will be easy to support and maintain, easily upgradeable, and will reap benefits for the employees, IT department, customers, and the company as a whole.

A wireless enterprise provides several solutions that lower the *total cost of ownership* (TCO), lower deployment cost, and with additional features that make it popular within the organization. A company can see a quick return on investment from a wireless enterprise deployment, with significant and almost immediate cost reductions in the cost of support and maintenance, specifically with mobile employees. With the wireless enterprise infrastructure in place, a company can offer a new level of convenience and ease-of-connectivity to mobile employees and provide new places to connect. These employees can be members of the sales team, employees who need to maintain mobility while connected to a corporate network, and even corporate guests. Within the office, whether it is an entire building consisting of multiple floors or just one floor in a large office building, companies can make the most of limited office space by employing a progressive office space model. Employees who visit the office infrequently from other locations or mobile employees will be able to work in an open space without the need of a dedicated cubicle or office. If designed and implemented properly, a *wireless local area network* (wireless LAN [WLAN]) can make networking in more than one area as simple as "power on and connect."

The term *wireless enterprise* can refer to several technologies or solutions or a combination of technologies and standards for wireless communications. A key component of building the wireless enterprise is the WLAN. A WLAN is a data communication system implemented as an extension or as an alternative for a wired LAN within an office, floor, building, or campus and home. WLANs transmit and receive data through the air, eliminating the need for wired physical connections to the corporate network. WLAN signals travel like normal radio signals; the signals do not require air or any other medium for transmission. They simply transmit information-using antennas.

WLANs have gained strong popularity in a number of vertical markets, including health care, retail, manufacturing, warehousing, and academia.

Today, WLANs are becoming more widely recognized as a connectivity alternative for a broad range of companies. Business Research Group, a market research firm, predicts a six-fold expansion of the worldwide WLAN market by the year 2004, reaching more than $4 billion in revenues.

In addition to providing solutions for the corporate workplace, WLAN technology is finding its way into the home. In a residence where recabling or cabling the premises is not an option, a WLAN is a quick, low cost alternative to cabling that offers the same benefits as a wired LAN. With an increasing number of households with multiple *personal computers* (PCs) and broadband connectivity, a WLAN enables sharing of resources, information, and Internet connectivity, and many of the same benefits as the corporate network. With the use of a broadband connection and a WLAN at home, the user will be able to establish a link with the corporate network and work without being restricted by the length of the network cable.

Establishing a standard and a best practices policy for your company early will allow the corporate user to connect wirelessly at the office and from home and provide low overhead administration. If a user has a mobile device, such as a laptop computer with a wireless network card, and the card is configured correctly, the user will be able to connect to the home network and the corporate network as easily as simply turning on the device. With a properly configured network card, a user can connect from home, come to the office, and connect without manually having to make changes, while keeping both network settings separate and secure. In companies with multiple offices, employees who rove from office to office can easily connect to the corporate WLAN easily and quickly.

What the Book Covers

As WLANs (wireless networks) grow, the new technologies continue expanding the reach of the corporate network to include new devices, new practices, and new places in which to conduct business. Devices, such as cellular phones, *personal digital assistants* (PDAs), laptop computers, and handheld computers, are creating a new means to communicate information and enabling greater accessibility to corporate applications, information, and resources. These new channels are connecting corporations to their customers and employees without the need of a wired connection to the corporate network infrastructure. In building a wireless enterprise, a good place to begin is by designing and implementing the WLAN.

Introduction

This book introduces the business benefits, uses, and technologies used for the implementation of WLANs. In a corporate environment, mobile users can access information and network resources as they attend meetings, collaborate with other users, or move to other buildings or campus locations.

After reviewing the business benefits and applications of WLANs, we will examine the various WLANs technologies that exist today. We will also discuss how these technologies are evolving and differ from each other and other components of the wireless enterprise.

We examine the basic components and technologies of WLANs and how they work together. We then explore the factors that customers must consider when evaluating WLANs for their business applications' needs. Lastly, Chapters 4, 5, and 6 provide key information on a how to design, implement, and support a WLAN for the office and for the home, including a step-by-step guide.

Should I Read This Book?

The decision to deploy a WLAN is rarely an individual choice. Although an individual may make the initial suggestion, it can require the efforts of several people and perhaps several teams of people across corporate departments to successfully architect, implement, and support the WLAN solution. Members of the IT staff, network administrators, network engineers, directors of technology, CTOs, CEOs, and CFOs will benefit from the information in this book. Provided are valuable decision-making and architectural tips and planning guides for the architecture, implementation, and support of the new infrastructure. The design guidelines are scalable enough to be used in large environments, corporate or academic campuses, or small organizations, home office users, or start-ups.

Applications of WLANs

Currently, WLANs *augment* rather than replace the existing wired LAN networks by providing the final few hundred feet of connectivity between a wired network backbone and the mobile user. For data center and server-to-server connectivity, wired networks continue to maintain a strong presence and are considered to be a best practice solution. In the case of such environments, moving to a wireless network does not offer a significant advantage over wired networks.

In certain areas, however, WLANs are providing a benefit over the wired networks and are having an immediate impact on the way people work and share data. The following list describes some of the many applications made possible through the power and flexibility of WLANs:

- *Doctors and nurses in hospitals* are more productive because handheld or notebook computers with WLAN capability deliver patient information instantly.

- *Consulting, accounting teams, or small workgroups* increase productivity with quick network setup and impromptu teams that can share information and proactively respond to changing business needs immediately. The need to contact a client's IT department for access and connectivity is greatly diminished, and the employee's efficiency is greatly increased.

- *Network managers in dynamic environments* minimize the overhead of moves, adds, and changes with WLANs, thereby reducing the total cost of LAN implementation and management. Making physical changes to the office, home, or anywhere else is no longer a monumental task.

- *Training sites at corporations and students at universities* use wireless connectivity to facilitate access to information and information exchanges and to enhance learning.

- *For some, working in the same environment for extended amounts of time* becomes distracting and inefficient. For them, being able to change the environment quickly and still maintain a network connection allows them to refresh their thinking process and focus on the task at hand.

- *Network managers installing networked computers in older buildings* find that WLANs are a cost-effective network infrastructure solution. The cost of running cabling to servers, cubicles, printing stations, and other network connected devices is a large chunk of the IT budget. Time and money can be saved by implementing a wireless network that makes connecting the network more cost effective.

- *Retail storeowners* use wireless networks to simply frequent network reconfiguration and provide network connectivity at the best place for the establishment. A wireless *point-of-sale* (POS) device can be placed at the best location without worry of wiring. Looking ahead, an employee with a portable wireless-enabled POS device can accept credit cards and decrease the wait in line.

- *Trade show and branch office workers* minimize setup requirements by installing preconfigured WLANs needing no local *management information system* (MIS) support. At trade shows, attendees and exhibitors benefit from using wireless networks to facilitate communication and data sharing.
- *Warehouse workers* utilize WLANs to exchange information with central databases and to increase their productivity. Large and small packages can be tracked and monitored to ensure storage in the correct location.
- *Network managers* can implement WLANs to provide backup for mission-critical applications running on wired networks.
- *Senior executives in conference rooms* make quicker decisions because they have real-time information at their fingertips for meetings and can assign tasks and follow up on preassigned items.

Wireless technology also provides benefits to other areas of an organization, as well as the IT department. Providing a WLAN will allow users to be more independent, adaptable, and independent. It will make a powerful business tool, the computer, more flexible and more powerful.

Wireless networks offer several business and technical benefits to the users and to network administrators. Once a wireless network is deployed, users will notice and appreciate the greater freedom that was not available with a traditional wired Ethernet network.

Before building the Wireless Enterprise, let us start by learning about the origins of wireless technology and the way it will integrate into an existing Ethernet network.

CHAPTER 1

The Wired Enterprise Background

Wireless local access networks (WLANs) are built upon the same network architecture and data transmission principle as current wired Ethernet networks. Whereas current wired networks use a physical connection to connect to the network, a WLAN can use *radio frequency* (RF) or *infrared* (IR) for the connection. The current emerging standard for wireless connectivity, 802.11b, uses RF and IR transmissions for communication (data transmission). Most vendors are providing wireless networking products that use RF for network connectivity versus IR. The distinct competitive advantage is that RF signals do not require line of sight for communication compared to an IR system. By using RF for the transmission layer, devices are able to connect through walls and floors up to several hundred feet without the need of a line-of-sight.

WLANs are designed to supplement and eventually replace traditional wired LANs. With an unobtrusive antenna that is relatively small, wireless networking devices eliminate the need for physical cables in an office, building, campus (educational or corporate), or

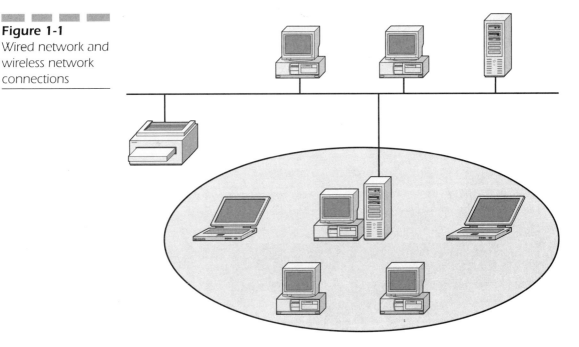

Figure 1-1
Wired network and wireless network connections

residence. A wireless network provides an excellent solution for a client-to-network connectivity, with the clients being computers, handhelds, and *personal digital assistants* (PDAs). Once on the network, the client can connect to other clients or servers, or connect outside the LAN to the Internet or other networks. WLANs based on the 802.11 standard are based on and built as an extension of the wired network architecture known as *Ethernet*. Figure 1-1 shows an example of a wireless network integrated with an existing wired Ethernet network.

Ethernet

Ethernet is a networking standard consisting of hardware and software used to connect computers, the hardware being a physical cable. Ethernet networks are not the same as the Internet, nor are the terms interchangeable. This distinction is important, as the history of Ethernet is not the same as the history of the Internet. In the simplest definition, the Internet is a collection of interconnected computers and interconnecting devices.

Ethernet is the common name for the *Institute of Electrical and Electronics Engineers* (IEEE) 802.3 series standard, based on the *Carrier Sense Multiple Access with Collision Detection* (CSMA/CD) access method that provides two or more stations with a means of sharing a common cabling system. A station can be a workstation, server, laptop, desktop computer, or a combination of other devices. CSMA/CD technology is the basis for Ethernet systems, which today range in speeds of 1, 10, 100, and 1,000 *megabits per second* (Mbps).

Known as a low-level network technology, Ethernet provides connectivity on the lower levels of the *Open Systems Interconnection* (OSI) model (see Figure 1-2). Ethernet supports IP and most other higher-level networking protocols. Traditional Ethernet, the early release, supports data transfers at rates of 1 Mbps and 10 Mbps. Over time, the performance needs of LANs have increased, and related technologies like Fast Ethernet (100 Mbps) and Gigabit Ethernet (1,000 Mbps) have been developed that extend traditional Ethernet speeds.

Figure 1-2
OSI model

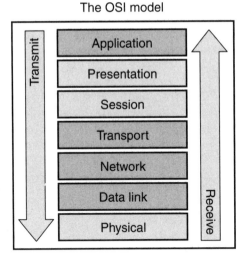

Higher-level network protocols like the *Internet Protocol* (IP) use Ethernet as their transmission medium. Data travels over the network in the form of frames, and collisions can occur when multiple stations on the network attempt to transmit simultaneously. The run length of Ethernet cables is limited to roughly 100 meters, but various networking devices (hubs, switches, and routers) allow the 100-meter limit to be passed. This makes Ethernet a cost-effective option for connecting computers located at varying distances from each other.

Ethernet has been a relatively inexpensive, reasonably fast, and very popular LAN technology for several decades. Competing technologies such as token ring came and went. Robert Metcalfe and D.R. Boggs of Xerox's *Palo Alto Research Center* (PARC) developed Ethernet in 1972, and specifications based on their work appeared in the IEEE 802.3 proposal in 1980. Ethernet specifications (IEEE 802.3) define low-level data transmission protocols and the technology needed to support them. In the OSI model, Ethernet technology exists at the physical and data link layers (layers 1 and 2). Figure 1-3 shows where 802.11 exists in the OSI Model.

The Wired Enterprise Background

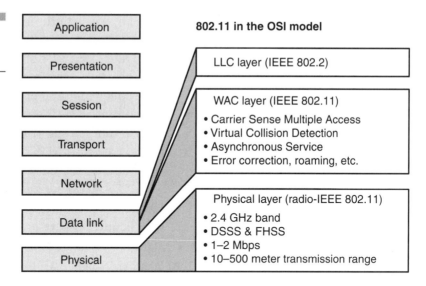

Figure 1-3
Physical layer and data link layer

What's in a Name?

In the early nineteenth century, the propagation of waves was a major preoccupation for British physicists. At that time, the physicists hypothesized that ether was a fluid medium through which electromagnetic waves propagated.

Robert Metcalfe chose to base the name of the technology on the word "ether" as a way of describing an essential feature of the system: The physical medium (that is, a cable) carries data bits, much the same way the old "luminiferous ether" was once thought to propagate electromagnetic waves through space.

On a minor note, the word Ethernet may be capitalized to signify the official 802.3 standard. It may also be used in the lower case to suggest a medium without switches, routers, and other intelligence.

Origins of Ethernet

The concept of Ethernet originally germinated in Robert Metcalfe's mind in 1970 as he read a paper by Norman Abramson of the

University of Hawaii from a computer conference that year. In the paper, Abramson told of another network design he called AlohaNet. AlohaNet was a packet-based radio data transmission system used for data communications among the Hawaiian Islands. AlohaNet was radio-based because of the sheer cost and challenge of providing cabling between all the islands made it an impractical solution.

At the time, AlohaNet allowed data to be transmitted wirelessly at 9,600 Kbps. The only problem was that AlohaNet offered no guarantees of packet delivery. Any client on the network could send packets to anyone else at any time. And to do so, a station simply began transmitting. If an acknowledgment was not received, the message had presumably failed to get through because packets had collided with each other, and in Metcalfe's words, "They were lost in the ether."

At that point, a station would simply wait for a random period of time to avoid a repeat collision as both stations returned to the channel at once and would retransmit the message. To Metcalfe, AlohaNet seemed a beautifully simple network. But Abramson showed that, because of collisions and other problems, it could use only 17 percent of its full capacity. That is, only 17 percent of the network carried new, useful data.

As a student of computer science looking for thesis ideas, Metcalfe believed that by using a form of advanced mathematics called *queuing theory* he could drastically improve the performance of AlohaNet without changing its essential elegance and simplicity. Later, as a graduate student at Harvard, what he eventually discovered would bring such a network toward 90 percent capacity. Overhead would be eliminated to 90 percent of packets to be useful data.

In 1972, while working at the now famous PARC, Metcalfe and his colleagues developed Ethernet. At PARC, some of the first personal computers were being made, and Metcalfe was asked to build a networking system for PARC's computers. Xerox was interested in building such a computer network because it was also building the world's first laser printer and wanted all of PARC's computers to be able to print to the new machine.

The press has often stated that Ethernet was invented on May 22, 1973, when Metcalfe wrote a memo stating the possibilities of its potential. Metcalfe asserts that Ethernet was actually invented gradually over a period of several years.

In the prophetic memo launching the concept, Metcalfe foreshadowed the secret of Ethernet's success and perhaps an eventual return to a wireless Ethernet. He wrote: "While we may end up using coaxial cable trees to carry our broadcast transmissions, it seems wise to talk in terms of an ether, rather than 'the cable' .... Who knows what other media will prove better than cable for a broadcast network: maybe radio or telephone circuits, or power wiring, or frequency-multiplexed cable TV or microwave environments, or even combinations thereof. The essential feature of our medium—the ether—is that it carries transmissions, propagates bits to all stations." In other words, it is the stations, rather than the network, that have to sort out and switch the messages. Looking back today, this is what makes Metcalf's idea a truly visionary statement.

In 1976, Metcalfe and Boggs (Metcalfe's assistant) published a paper titled, "Ethernet: Distributed Packet-Switching For Local Computer Networks." Three years later, Metcalfe left Xerox to promote the use of personal computers and LANs. He successfully convinced Digital Equipment, Intel, and Xerox Corporations to work together to promote Ethernet as a standard.

In 1983, Novell created a proprietary Ethernet frame type prior to the release of the IEEE 802.3 specification for use with their networking software. By 1985, the IEEE 802.3 specification was completed and provided a specification for Ethernet connectivity over thick coaxial (coax) cable and thin coaxial cable. In 1990, the specification was updated to include Ethernet over twisted-pair copper wiring. The current IEEE 802.3 specification includes thick coax, thin coax, twisted-pair cabling, and fiber, with speeds of 10, 100, and 1,000 Mbps. Now an international computer industry standard, Ethernet is the most widely installed LAN protocol.

Traditional Wired Ethernet

Often referred to as Thicknet, 10Base5 technology was the first incarnation of Ethernet and used a thick coaxial cable for network connectivity (see Figure 1-4). It was used in the 1980s until 10Base2, or Thinnet, appeared with more flexible cabling. Both the 10Base5 and 10Base2 technologies used coaxial cable as the media for communication. The most common form of traditional Ethernet, however, is

Figure 1-4
Coaxial cable

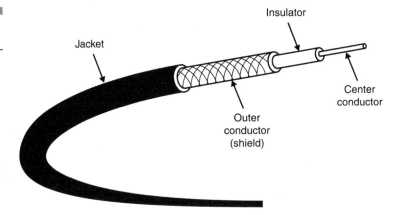

Table 1-1
Early wired Ethernet standards and cable

Name	Segment Length (Max.)	Cable
10Base5	500 m/1640 ft	RG-8 or RG-11 coaxial
10Base2	185 m/606 ft	RG 58 A/U or RG 58 C/U coaxial
10Base-T	100 m/328 ft	Category 3 or better UTP

10Base-T because of the inherent advantages of *unshielded twisted pair* (UTP) over coaxial cabling and its low cost compared to alternatives like fiber.

Several other less well-known Ethernet standards exist, including 10Base-FL, 10Base-FB, and 10Base-FP for networks that use a fiber optic cable, as well as 10Broad36 for broadband *cable television* (CATV) cabling.

Besides the type of cable involved, another important factor in Ethernet networking is the segment length. A single uninterrupted network cable can only span a certain physical distance before its electrical characteristics are critically affected by factors such as line noise or reduced signal strength. Table 1-1 lists these well-known forms of Ethernet technology and their basic characteristics.

The Wired Enterprise Background

Fast Wired Ethernet

In the mid-1990s, Fast Ethernet achieved its design goal of improving traditional Ethernet's performance while avoiding the need to completely recable existing networks. Fast Ethernet comes in two major varieties:

- 100Base-T (using UTP cable)
- 100Base-FX (using fiber optic cable)

By far, the most popular of these is 100Base-T, a standard that includes 100Base-TX (Category 5 UTP), 100Base-T2 (Category 3 or better UTP), and 100Base-T4 (100Base-T2 cabling modified to include two additional wire pairs). The most obvious reason for the advantage is cost. Copper twisted-pair cable continues to be a better value and is cheaper to install than fiber optic cable. Figure 1-5 shows a typical UTP cable with four pairs of wires.

Gigabit Ethernet

Whereas Fast Ethernet improved traditional Ethernet from 10 to 100 Mbps speed, Gigabit Ethernet offers the same order-of-magnitude improvement over Fast Ethernet by offering speeds of 1,000 Mbps (1 gigabit per second). Gigabit Ethernet was first made to travel over optical and copper cabling, but the 1000Base-T standard successfully supports it as well. 1000Base-T uses Category 5 cabling similar to 100 Mbps Ethernet, although achieving gigabit speed requires the use of additional wire pairs.

Figure 1-5
Category 5 UTP cable

Different Physical Ethernet Network Types

The Base in the various Ethernet network topologies refers to *baseband*. A baseband network has a single channel that is used for communication between stations. Ethernet specifications that use Base in the name refer to baseband networks.

A broadband network is much like CATV, where different services communicate across different frequencies on the same cable. Broadband communications allow an Ethernet network to share the same physical cable as voice or video services. 10Broad36 is an example of broadband networking. The frequency of the data transmission on the broadband network cable, fiber, or twisted pair does not change.

Some of the physical Ethernet types as defined in the 802.3 specification are as follows:

- ***1Base5*** A specification of Ethernet that runs at 1 Mbps over twisted-pair wiring. This physical topology uses centralized hubs or switches to connect network devices.

- ***10Base2*** The original design for a departmental or workgroup-sized Ethernet environment. It is designed to be simple, inexpensive, and flexible as people and stations move. It is designed as a smaller and less expensive alternative to 10Base5 and is sometimes referred to as Thinnet or thin Ethernet because of the much smaller cables. 10Base2 is a bus topology, but each of the workstations use a T *British Naval Connector* (BNC) to connect workstations to the central bus.

- ***10Base5*** The original Ethernet backbone, 10Base5 is occasionally referred to as Thicknet or thick Ethernet because of the thick, 50 ohm coaxial cable used as the physical medium. It was originally designed to be used with the traditional Ethernet backbone and to be left in place permanently or for extended periods. It is a bus topology that uses transceiver cables to attach stations to the central 10Base5 cable.

- ***10Base-F*** A set of optical-fiber-medium specifications that defines connectivity between devices.

- **10Base-T**   This provides Ethernet services over twisted-pair copper wire. It offers connectivity at speeds of 10 Mbps and can be used to connect 2 computers through a hub.
- **10Broad36**   A seldom-used Ethernet specification that uses a physical medium similar to CATV, with CATV type cables, taps, connectors, and amplifiers.
- **100Base-T**   A series of specifications that provides 100 Mbps speeds over copper or fiber and is often referred to as Fast Ethernet.
- **Fiber Optic Inter-Repeater Link (FOIRL)**   A specification of the 802.3 standard that defines a standard means of connecting Ethernet repeaters via optical fiber.
- **Gigabit Ethernet**   Provides speeds of 1,000 Mbps over copper and fiber.

Present-Day Ethernet Networks

Most modern Ethernet networks use twisted-pair copper cabling or fiber to attach devices to the network. Modern Ethernet implementations, those networks deployed around early 1990, often look nothing like the traditional deployments. Although legacy Ethernet networks transmitted data at 10 Mbps and lower, modern networks can operate at 100 or even 1,000 Mbps. Modern Ethernet networks use twisted-pair wiring or fiber optics to connect stations in a radial pattern known as a star topology. In the past, long runs of coaxial cable provided attachments for multiple stations in a bus topology.

Bus Topology   A *bus topology* is a networking architecture that is linear, usually created by using one or more pieces of cable to form a single line or bus (see Figure 1-6). The signals sent by one station extend the length of this cable to be heard by other stations. Bus networks (not to be confused with the system bus of a computer) use a common backbone to connect all devices. A single cable functions as the backbone and a shared communication medium that devices attach or tap into with an interface connector. A device wanting to

Figure 1-6
Bus topology

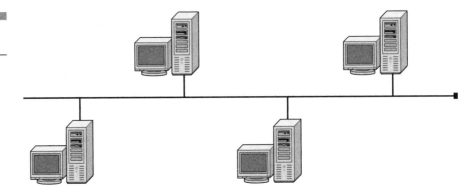

communicate with another device on the network sends a broadcast message onto the wire that all other devices see, but only the intended recipient accepts and processes the message. Information about the station that is the intended recipient is contained in the data packet, along with the actual data the station is supposed to receive.

Star Topology A *star topology* is a networking architecture that includes a central device or hub that connects all stations together. Signals sent by a station must pass through central hubs. Since the hub sits in the center and all other stations are linked through the hub, the architecture resembles a star or a bicycle hub with spokes emanating from the center (see Figure 1-7). Most modern Ethernet networks use the star topology. A star network features a central connection point that may be an actual hub or a switch for transmitting data. Devices typically connect to the hub with UTP cable.

Compared to the bus topology, a star network generally requires more cable, but a failure in any network segment will only take down network access for that segment and not the entire network. If the hub fails, however, the entire network also fails. The star topology became popular quickly and early because of the ease in troubleshooting and implementing this architecture. Segments of the network could be thoroughly tested and deployed individually, resulting in a reliable network infrastructure. During deployment,

The Wired Enterprise Background

Figure 1-7
Star topology

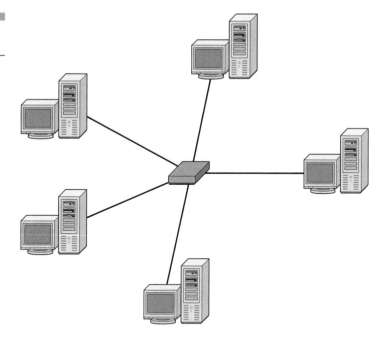

Figure 1-8
Modern Ethernet network

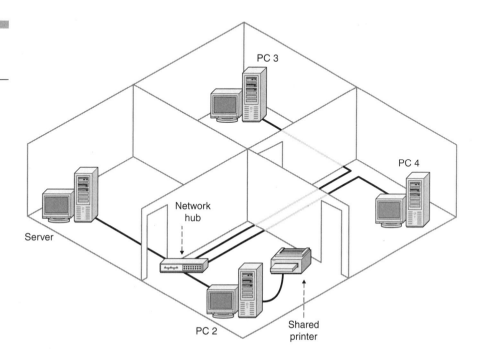

segments of the network could be live, without having to wait for the entire network to be available. As a result, high-priority segments could be brought live first to minimize downtime.

Perhaps the most striking advancement in contemporary Ethernet networks is the use of *switched Ethernet*. Switched networks replace the shared medium of legacy Ethernet with a dedicated segment for each station. These segments connect to a switch, which acts much like an Ethernet bridge, but can connect many of these single station segments. Some switches today can support hundreds of dedicated segments.

Since the only devices on the segments are the switch and the end station, the switch picks up every transmission before it reaches another station. The switch then forwards the data packet over the appropriate segment, just like a bridge. Because any segment contains only a single node, the data packet only reaches the intended recipient. This allows many conversations to occur simultaneously on a switched network with few collisions.

Ethernet switching has given rise to another advancement: *full-duplex* Ethernet. Full-duplex is a data communications term that refers to the capability to both send and receive data at the same time. Legacy Ethernet is half-duplex, meaning a device on the network can transmit or receive at any given time. In a totally switched network, nodes only communicate with the switch and never directly with each other. Switched networks also employ either twisted-pair or fiber optic cabling, both of which use separate conductors for sending and receiving data. In this type of environment, Ethernet stations can forgo the collision detection process and transmit at will, since they are the only potential devices that can access the medium. This allows end stations to transmit to the switch at the same time it transmits to them, achieving a collision-free environment.

802.3

You may have heard the term 802.3 used in place of or in conjunction with the term Ethernet. Ethernet originally referred to a networking implementation standardized by Digital, Intel, and Xerox. (For this reason, it is also known as the DIX standard.)

In February of 1980, the IEEE created a committee to standardize network technologies. The IEEE titled this the 802 working group, named after the year and month of its formation. Subcommittees of the 802 working group separately addressed different aspects of networking. The IEEE distinguished each subcommittee by numbering it 802.X, with X representing a unique number for each subcommittee. The 802.3 group standardized the operation of a CSMA/CD network that was functionally equivalent to the DIX Ethernet.

Ethernet and 802.3 differ slightly in their terminology and the data format for their frames, but are in most respects identical. Today the term Ethernet refers generically to both the DIX Ethernet implementation and the IEEE 802.3 standard.

Token Ring In the early 1990s, the most common LAN alternative to Ethernet was a network technology developed by IBM called token ring. Although Ethernet relies on the random gaps between transmissions to regulate access to the medium, token ring implements a strict, orderly access method. A token ring network arranges nodes in a logical ring (see Figure 1-9). The nodes forward frames in one direction around the ring, removing a frame when it has circled the ring once. The ring initializes by creating a token, which is a special type of data frame that gives a station permission to transmit. The token circles the ring like any frame until it encounters

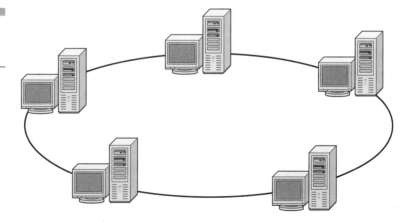

Figure 1-9
Small token ring network

a station that wishes to transmit data. This station then captures the token by replacing the token frame with a data-carrying frame, which encircles the network. Once that data frame returns to the transmitting station, that station removes the data frame, creates a new token, and forwards that token on to the next node in the ring.

Token ring nodes do not look for a carrier signal or listen for collisions; the presence of the token frame provides assurance that the station can transmit a data frame without fear of another station interrupting. Because a station transmits only a single data frame before passing the token along, each station on the ring will get a turn to communicate in a deterministic and fair manner. Token ring networks typically transmit data at either 4 or 16 Mbps.

Fiber Distributed Data Interface (FDDI) is a token-passing technology that operates over a pair of fiber optic rings, with each ring passing a token in opposite directions. FDDI networks offer transmission speeds of 100 Mbps, which initially made them quite popular for high-speed networking. With the advent of 100 Mbps Ethernet, which is cheaper and easier to administer, FDDI has waned in popularity as a practical solution for companies.

Asynchronous Transfer Mode (ATM) A final network technology that bears mentioning is ATM. ATM networks blur the line between local and wide area networking, being able to attach many different devices with high reliability and at high speeds, even across the country. ATM networks are suitable for carrying not only data, but voice and video traffic as well, making them versatile and expandable. Although ATM has not gained acceptance as rapidly as originally predicted, it is nonetheless a solid network technology for the future.

Which Ethernet Topologies Are No Longer Popular? A number of physical networking components are detailed in the IEEE 802.3 specification, but many of those early physical networking components are not used in most modern Ethernet networks. However, there may be instances where an existing legacy network still uses these older components. Since these older pieces of equipment are still part of the 802.3 specification, there are no technical reasons why an Ethernet network would not operate properly with these components. The two most popular older Ethernet technologies are 10Base5 and 10Base2.

Ethernet in Use

An Ethernet network functions nearly identically to other networks of similar size no matter its speed or layout. Devices plugged into the network possess a *network interface card* (NIC), more generally, a network adapter or network card that interfaces directly to the computer system bus. The NIC includes a cable connector such as the RJ-45 connector used with UTP cable connections or a BNC connector used with coaxial cable connections. The original versions of Ethernet used very different connectors.

Data sent over the Ethernet exists in the form of frames. An Ethernet frame contains two headers and a data section having a combined length of no more than 1,518 bytes. The standard calls for frames to be broadcast to all devices, meaning that network adapters must explicitly recognize and discard all frames that are not intended for the device to receive.

Devices wanting to transmit on the Ethernet first perform a very fast check to determine whether the medium is available or whether a transmission is currently in progress. If the Ethernet is available, the device transmits. However, the Ethernet standard does not prevent multiple devices from transmitting simultaneously. When that happens, these collisions cause both transmissions to fail and require both devices to retransmit at a later time. The devices then wait a random time interval to retry a transmission.

Because of its simplicity and elegance of design, Ethernet has not changed a great deal over the years. What *is* changing is the medium used in connecting the network devices; the medium is not a cable. Ethernet has at long last returned to its original roots of the old "luminiferous ether" and of AlohaNet. Ethernet network communication is starting to occur through air, replacing the earlier cable connections.

Wireless networking is combining the functionality and freedom of AlohaNet, with the reliability of Ethernet. By taking a step back to AlohaNet, Ethernet is able to take a giant leap forward to wireless Ethernet.

WLAN Advantages

The widespread reliance on networking in business and the meteoric growth of the Internet and online services are strong testimonies to the benefits of shared data and resources. With WLANs, users can access shared information without looking for a place to plug in, and network managers can set up or augment networks without installing or moving wires. WLANs offer productivity, convenience, and cost advantages over traditional wired networks. Once connected to the corporate WLAN, users can access all the resources that users connecting from a wired LAN would access, including the Internet. In a WLAN, only the method of connectivity changes, the physical layer and the data link layer on the OSI model. To the user, this change is transparent.

The second branch of wireless networking involves using *third-generation* (3G) technologies to access data. Whereas WLANs provide laptops, desktops, and other computing devices with access to the corporate network, 3G allows a whole new set of devices to access data. Some of these new devices already exist and more are coming on the market. These new devices promise to provide access to the Internet and to multimedia. What remains is access to corporate data, critical to businesses whose goal is to provide their employees with access to corporate data.

WLANs offer the following productivity, service, convenience, and cost advantages over traditional wired networks:

- ***Mobility*** WLAN systems can provide LAN users with access to real-time information anywhere in their organization. This mobility supports productivity and service opportunities not possible with wired networks.

- ***Installation speed and simplicity*** Installing a WLAN system can be fast and easy, and can eliminate the need for installing cable through walls and ceilings. Thus, it can result in a tremendous time savings in the initial deployment.

- ***Cost*** A WLAN implementation includes both infrastructure costs for the wireless access points and user costs for the WLAN adapters. Infrastructure costs depend primarily on the number of access points deployed; access points range in price from $200 to $2,000. The number of access points typically depends on the required coverage region and/or the number and type of users to be serviced. WLAN adapters are required for the computers and range in price from $150 to $500.

- ***Reduced total cost of ownership (TCO)*** The long-term cost benefits are greatest in dynamic environments requiring frequent moves, adds, and changes. This solution provides the most benefits in growing companies and in start-up or small businesses. Because WLANs simplify moves, adds, and changes of new clients to the network, they reduce the TCO of user downtime and administrative overhead. A WLAN also eliminates the direct costs of cabling and the labor associated with troubleshooting and repairing it.

- ***Installation flexibility*** Wireless technology allows the network to go where wire cannot go quickly and cheaply. In one example, manufacturing and warehousing buildings use WLANs to monitor inventory and track shipments.

- ***Licensing issues*** In the United States, the *Federal Communications Commission* (FCC) governs radio transmissions, including those employed in WLANs. Other nations have corresponding regulatory agencies. WLANs are typically designed to operate in portions of the radio spectrum where the FCC does not require the end user to purchase a license to use the airwaves. In the United States, most WLANs broadcast over one of the *Instrumentation, Scientific, and Medical* (ISM) bands. These include 902 to 928 MHz, 2.4 to 2.483 GHz, 5.15 to 5.35 GHz, and 5.725 to 5.875 GHz. For WLANs to be sold in a particular country, the manufacturer of the WLAN must ensure its certification by the appropriate agency in that country. This responsibility falls on the manufacturer of the WLAN product and not on the company purchasing the WLAN.

- ***Simplicity and ease of use*** Users need very little new information to take advantage of WLANs. Because the wireless nature of a WLAN is transparent to a user's *operating system* (OS), applications work the same as they do on wired LANs. This means that users will not require additional training or need to learn a new technology.

WLANs simplify many of the installation and configuration issues that plague network managers. Finally, the portable nature of WLANs lets network managers preconfigure and troubleshoot entire networks before installing them at remote locations. Once configured, WLANs can be moved from place to place with minimal or no modification.

Technical Advantages of WLANs

In addition to business advantages, WLANs also offer several technical advantages. The following are the most compelling technical advantages of WLANs:

- ***Scalability*** WLAN systems can be configured in a variety of topologies to meet the needs of specific applications and installations. Configurations are easily changed and range from peer-to-peer networks suitable for a small number of users to full infrastructure networks for thousands of users that allow roaming over a broad area, such as a corporate or college campus.
- ***Battery life for mobile platforms*** End-user wireless network cards are designed to run off the AC or battery power from their host notebook or hand-held computer, since they have no direct wire connectivity of their own. WLAN vendors typically employ special design techniques to maximize the host computer's energy usage and battery life. This ensures that a WLAN network card and wired network cards will use a comparable amount of power.
- ***Safety*** The output power of WLAN systems is very low, much less than that of a hand-held cellular phone. Since radio waves fade rapidly over distance, those in the area of a WLAN system

receive very little exposure to the RF energy. As a result, users are not exposed to a damaging amount of RF leverage. WLANs must also meet stringent government and industry regulations for safety. No adverse health affects have ever been attributed to WLANs.

- ***Installation flexibility***   Wireless technology allows the network to go where wire cannot. Wireless networks can be designed to be extremely simple or quite complex. Wireless networks can support large numbers of nodes and/or large physical areas by adding access points to boost or extend coverage.

- ***Compatibility with existing networks***   Today's WLANs provide for industry-standard interconnection with wired networks such as Ethernet or token ring. WLAN nodes are supported by network OSs in the same fashion as any other LAN node: through the use of the appropriate drivers. Once installed, the network treats wireless nodes like any other network component.

- ***Interference and coexistence***   The unlicensed nature of radio-based WLANs means that other products that transmit energy in the same frequency spectrum can potentially provide some measure of interference to a WLAN system. Microwave ovens are a potential concern, but most WLAN products are designed to account for microwave interference.

- ***Simplicity and ease of use***   WLAN products incorporate a variety of diagnostic tools to address issues associated with the wireless elements of the system; however, products are designed for administrators, so most users rarely see a need to use these tools.

Flexibility and mobility make WLANs both effective extensions and attractive alternatives to wired networks. WLANs provide all the functionality of wired LANs without the physical constraints of the wire itself. WLAN configurations range from simple peer-to-peer topologies to complex networks offering distributed data connectivity and roaming. Besides offering end-user mobility within a networked environment, WLANs enable portable networks, allowing LANs to move with the knowledge workers who use them.

Summary

WLANs provide technical and business advantages. They enable an enterprise, business, or academic institution to implement a flexible and robust solution over a wide area. WLANs meet the business needs and technical requirements of businesses and provide additional advantages over the wired LAN. The biggest gain for some may be the freedom from a tether.

CHAPTER 2

Welcome to the Wireless Enterprise

When Xerox completed the Ethernet project in the early 1970s, the face of computing was changed forever. Now users could share and access resources at fairly high speeds, as measured by 1970s standards anyway.

However, there continued to be limitations to this wonderful technology. The workstation had to be tethered to the network by a wire. Just as today, laptops at the time were not as mobile because being connected with a wire took all the mobility out of the picture, especially if you purchased a laptop or mobile workstation for the advantage of mobility. In order to prepare an office for a *local area network* (LAN), the walls had to be ripped open to run the cabling. Or there were unsightly cables running along the walls and desks, which created a hazard and an eyesore in the office environment. In many instances, this cabling solution was not a viable option. Older buildings could not be ripped to run cabling because of asbestos cleanup and removal. In buildings with no pre-existing network cabling, the time and cost of laying in cabling were too extensive and required possible permissions from the leaser or landlord. For most businesses, the cost of cabling the office was a huge chunk of the budget. It just was not very economical.

In contrast to computers, there have been cordless communication devices in most aspects of our lives: cordless phones, cellular phones, pagers, and radio transmitters and receivers. Some of these communication devices, like cordless phones, provide coverage over a limited area, such as a home or office. Other devices, like cell phones, provide coverage over a much larger area, such as a city or the entire nation. Most of these devices transmit on frequencies that do not require a license and are proprietary offerings. Each company does things its own way because these devices do not have to interact with other manufactures' devices. In the early days of these technologies, a standard was not in place because one was not required, and each device was not required to interact with each another.

In computing, however, because of the sheer expanse of the market, a standard is required. After years of development and fighting over the creation of a standard for wireless connectivity to the Internet, a standard for wireless data transmission has emerged. All wireless networking products adhere to the 802.11 standard, an evolution of the 802.3 network standard. The 802.11 standard continues to evolve; to date, there are 802.11a, 802.11b, and others that are improvements on the original 802.11. All the while, the technology is improving, making it easier and cheaper for companies to go wireless while offering connections at speeds comparable to Ethernet.

Much like the early Ethernet offerings, the first *wireless LAN* (WLAN) technologies offered around 1995 were low-speed (1 to 2 Mbps) proprietary offerings. Despite these shortcomings, the freedom and flexibility of wireless allowed these early products to find a place in vertical markets, such as retail and warehousing, where mobile workers use handheld devices for inventory management and data collection. Later, hospitals applied wireless technology to deliver patient information right to the bedside. As computers made their way into the classrooms, schools and universities began installing wireless networks to avoid cabling costs and share Internet access.

The pioneering wireless vendors soon realized that for the technology to gain broad market acceptance, an Ethernet-like standard was needed. The vendors joined together in 1991, first proposing, and then building, a standard based on contributed technologies. In June 1997, the *Institute of Electrical and Electronics Engineers* (IEEE) released the 802.11 standard for wireless local area networking.

With the advent of the 802.11 wireless standards, many *information technologies* (IT) administrators realized the headaches associated with wiring up an office were over. Although the standard was finalized in 1997, it was not until early 2000 that WLAN products made it into the mainstream networking marketplace. Today a cornucopia of devices and technologies are available for wireless networks. Today's most popular wireless networking products use *radio frequency* (RF) technology to transmit and receive data over the air, eliminating the need for wired connections.

WLANs combine data connectivity with user mobility. For the corporate professional, keeping information current on a *personal digital assistant* (PDA), a laptop, or a home or work computer is highly important, and a wireless network makes connecting the various devices effortless. The primary function of this new technology is to connect the user with the network. For data centers and server-to-server communication, wired Ethernet is still the most widely used option.

For some time now, companies and individuals have interconnected computers with LANs. This has enabled users to access and share data, applications, and other services not resident on any one computer. LAN users have at their disposal much more information, data, and applications than they could otherwise store by themselves. In the past, all LANs were wired together and in a fixed location (see Figure 2-1).

There are many reasons for deploying a WLAN in an enterprise. An increasing number of LAN users are becoming mobile. These mobile users require that they be connected to the network regardless of where they are because they want simultaneous access to the network. This makes the use of cables, or wired LANs, impractical if not impossible. WLANs are very easy to implement because one access point can serve a large area. There is no longer a need for wiring every workstation and every room. If a workstation must be moved, it can be done easily and without additional wiring, cable drops, or reconfiguration of the network. Figure 2-2 shows an example of several devices connected to a WLAN.

Another advantage is its portability. If a company moves to a new location, the wireless system is much easier to move than ripping up all the cables that a wired system would have throughout the building and then attempting to rewire the new location. Most of these advantages translate into monetary and time savings.

Returning to AlohaNet

WLANs have gained popularity only recently, but the history of wireless networking stretches farther back than you might think. This type of wireless data transmission technology was first introduced during World War II when the Army began sending battle plans over enemy lines and when Navy ships instructed their fleets via ship-to-shore communication. The military configured wireless signals to transmit data over a medium that had complex encryption, which made unauthorized access to network traffic virtually impossible. Wireless technology eventually progressed as an invaluable tool used by the U.S. military.

Welcome to the Wireless Enterprise

Figure 2-1
Traditional wired LAN

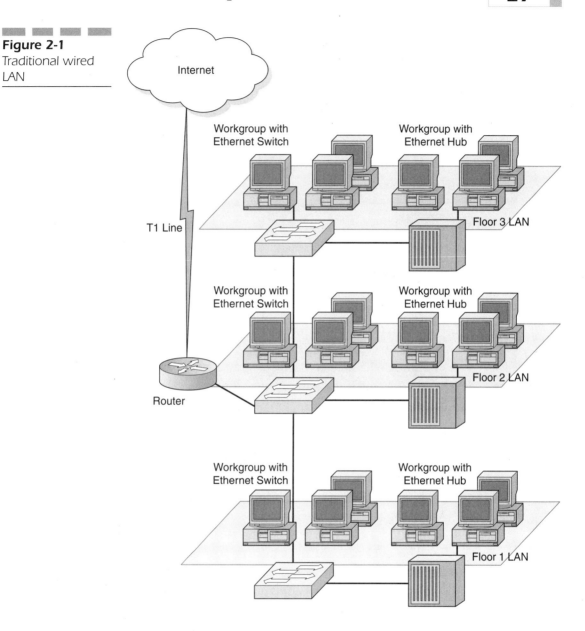

Learning about this technology inspired a group of researchers at the University of Hawaii to create the first packet-based radio communications network in 1971. AlohaNet, as it was named, was essentially the very first wireless network, although its development took a long period of time. This first *wireless wide area*

Figure 2-2
WLAN

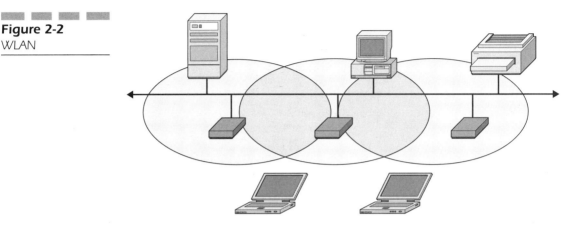

network (WWAN) consisted of seven computers that communicated in a bidirectional star topology that spanned four of the Hawaiian Islands, with the central computer based on Oahu Island.

The AlohaNet developed by Norman Abramson went into operation in 1971 and was the first-ever packet radio network, operating at 9,600 *bits per second* (bps) throughout the state of Hawaii. Aloha is also the name of the protocol for satellite and terrestrial radio transmissions. It allows stations to transmit at any time but risks collisions with other stations if more than one station transmits at the same time. Slotted Aloha reduces the chance of collisions by dividing the channel into time slots and requiring the station to send only at the beginning of a time slot.

The evolution of networking started with AlohaNet and took a giant leap with Ethernet. Ethernet was more efficient and considerably faster, but it required the use of wires as the connecting medium. The third incarnation, wireless networking, combines the reliability, speed, and simplicity of Ethernet with the advantages of a wireless data network, originated by AlohaNet.

Wireless technology proved so valuable as a secure communications medium that many businesses and schools thought it could increase their computing arena by expanding their LAN through the use of WLANs. And so, wireless technology as we know it began its journey in the 1970s to arrive into every house, classroom, and business around the world in the 1990s.

The first WLAN technologies for commercial use were offered around early 1994, but were low-speed (1 to 2 Mbps) proprietary offerings that operated in the 900 MHz band. As a result, the early adopters of wireless networking technology were limited to one vendor and to speeds that were one-tenth the speed of wired Ethernet networks. Figure 2-3 shows the EM spectrum and the various devices that share it. Today's WLANs operate in the 2 GHz range.

Just as the 802.3 Ethernet standard allows for data transmission over twisted pair and coaxial cable, the 802.11 WLAN standard allows for transmissions over different media. Compliant media includes *infrared* (IR) light and two types of radio transmission technology within the unlicensed 2.4 GHz frequency band: *frequency hopping spread spectrum* (FHSS) and *direct sequence spread spectrum* (DSSS).

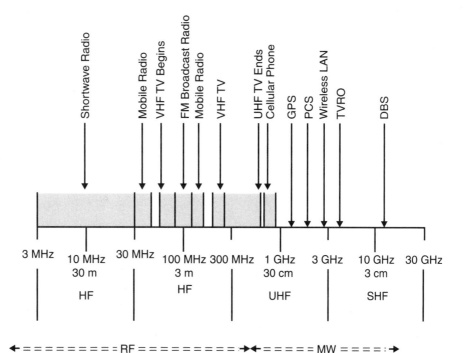

Figure 2-3
Electromagnetic spectrum

Spread spectrum is a modulation technique developed in the 1940s that spreads a transmission signal over a broad band of radio frequencies. This technique is ideal for data communications because it is less susceptible to radio noise and creates little interference.

FHSS is limited to a 2 Mbps data transfer rate and is recommended for only very specific applications such as certain types of watercraft. For all other WLAN applications, DSSS is the better choice. The recently released evolution of the IEEE standard, 802.11b, provides for a full Ethernet-like data rate of 11 Mbps over DSSS, as FHSS does not support data rates greater than 2 Mbps.

Today's WLAN Technology

In 1990, the IEEE 802 Executive Committee established the 802.11 working group to create a WLAN standard. The standard specified an operating frequency in the 2.4 GHz *Industrial, Scientific, and Medical* (ISM) band and began laying the groundwork for a cutting-edge technology.

Seven years later, the group approved IEEE 802.11 as the world's first WLAN standard with data rates of 1 and 2 Mbps. The 802.11 standard provided 1 or 2 Mbps transmission in the 2.4 GHz band using either a FHSS technique or DSSS, also known as *Code Division Multiple Access* (CDMA).

The 802.11 Standard

The Executive Committee anticipated the need for a more robust and faster technology to provide connectivity. Almost immediately, the committee began work on another 802.11 extension that would satisfy the future needs of greater bandwidth and broad acceptance. Within 24 months, the working group approved two *Project Authorization Requests* (PARs) for higher-rate, physical-layer extensions to 802.11. The two extensions were designed to work with the existing 802.11 *Medium Access Control* (MAC) layer, with one being the IEEE 802.11a for 5 GHz and the other IEEE 802.11b for 2.4 GHz.

The 802.11b standard is much like the 802.11 standard; it supports transmission in IR light and two types of radio transmission within the unlicensed 2.4 GHz frequency band: FHSS and DSSS.

802.11b

The IEEE 802.11b standard is an expansion of the IEEE 802.11 standard that allows transmission speeds of up to 11 Mbps at distances of several hundred feet. The distance depends on impediments, materials, environment, and line of sight for IR-based networks.

This specification started to appear in commercial form in mid-1999, with Apple Computer's introduction of its AirPort components, manufactured in conjunction with Lucent's WaveLAN division. The standard is backwards compatible with earlier specifications (802.11), enabling speeds of 1, 2, 5.5, and 11 Mbps on the same transmitters using DSSS for transmission.

The DSSS radio system works on a fixed, preconfigured channel. Although this approach is required in order to support higher bandwidth, it does make the system sensitive to interference from other radio signals using the same frequency. This is the standard version used by the 11 Mbps DSSS system offered by various vendors.

It is possible to have three access points with three different, noninterfering, nonoverlapping channels in one physical location without any radio planning. In larger networks using more than three access points within the same coverage area, the DSSS products require radio planning in order to achieve an optimum performance at higher data speeds. Overlapping frequencies may result in performance speeds lower than 11 Mbps, and a fourth access point introduced into the same coverage area will affect the capacity of the entire wireless network.

The actual data rate depends on the quality of the received signal and the interference from additional overlapping cells using the same radio channel. This may result in less than 100 percent of the coverage area being provided with full 11 Mbps coverage. Depending on the received signal's strength, which is related to the distance between an access point and the user's terminal as well as to the level of interference, the system adapts the data rate from 11 Mbps during optimal conditions to 5.5 Mbps, 2 Mbps, and as low as 1 Mbps in more adverse conditions.

802.11a

In place of 802.11b's direct sequence, 802.11a uses *coded orthogonal frequency division multiplexing* (COFDM). This technology was built from the ground up by breaking down the 20 MHz high-speed data channel into 52 lower-speed subchannels that are sent in parallel. Forty-eight of these subchannels are for data, with the remaining four reserved for error correction. These subcarriers are divided so that each one of them are orthogonal to each other, hence the name, thus allowing them to be packed together much closer than standard frequency division multiplexing. It is due to this that OFDM can provide the superior spectral efficiency.

By operating at a higher frequency in the microwave range, 802.11a gains speed. The cost of more bandwidth is less range, and 802.11a tries to beat the distance problem by using more efficient data encoding schemes and by increasing the power used to deliver its signal to a high, by WLAN standards, 50 *milliwatts* (mW).

As defined by 802.11, 802.11a operates in the 5 GHz *Unlicensed National Information Infrastructure* (UNII) band. This band provides 300 MHz of bandwidth, with the first 200 MHz residing between 5.15 and 5.35 GHz, and the last 100 MHz between 5.725 and 5.825 GHz. In the first 100 MHz frequency range, there is a restriction of 50 mW as the maximum power output. For the second and third frequency ranges, the maximum power output is limited to 250 mW and 1 W, respectively. Although 802.11a and 802.11b have relatively identical MAC layers, 802.11a uses a different physical layer.

802.11a is analogous to the Fast Ethernet of Ethernet and can reach speeds of 54 Mbps, compared to 802.11b's 11 Mbps. Because the physical layer shows such different characteristics, 802.11b devices are completely incompatible with 802.11a.

In 2000, IEEE established a standard for another type of network, the wireless *personal area network* (PAN).

PAN Technology

PANs are designed to give devices close to each other connectivitity without the need for wires. The emerging standard for this technology is called *Bluetooth*.

Bluetooth

A *wireless PAN* (WPAN) technology from the Bluetooth Special Interest Group, Bluetooth was founded in 1998 by Ericsson, IBM, Intel, Nokia, and Toshiba. Bluetooth is an open standard for the short-range transmission of digital voice and data between mobile devices (laptops, PDAs, and phones) and desktop devices. It supports point-to-point and multipoint applications.

The name Bluetooth is taken from King Harald Blatan (known as Bluetooth) of Denmark who in the tenth century began to Christianize the country. Ericsson, a Scandinavian company, was the first to develop this specification and chose this name.

Bluetooth provides data transfers of up to 720 Kbps within a range of 10 meters and up to 100 meters with a power boost. Unlike the *Infrared Data Association* (IrDA), which requires that devices be aimed at each other (line of sight), Bluetooth uses radio waves that can transmit through walls and other nonmetal barriers. Bluetooth transmits in the unlicensed 2.4 GHz band and uses a FHSS technique that changes its signal 1,600 times per second. If there is interference from other devices, the transmission does not stop, but its speed is downgraded. Bluetooth works in the same RF range as the 802.11 standard, and although both have taken abuse from critics, the two technologies are designed to fulfill different needs and can coexist in the same environment.

In spite of the criticisms, Bluetooth technology is an upcoming WPAN technology that has gained significant industry support and will coexist with most WLAN solutions. The specification is for a small, form-factor, low-cost radio solution that can provide mobile phones, mobile computers, and other portable handheld devices with connectivity to the Internet. This technology, embedded in a wide range of devices to enable simple, spontaneous wireless connectivity, is a complement to WLANs, which are designed to provide continuous connectivity via standard wired LAN features and functionality.

The 802.11 standard gives computers and laptops the capability to connect to the corporate network and beyond: to the Internet. Bluetooth allows various devices, such as cell phones, pagers, and computers, to connect with each other in an ad hoc network.

Building-to-Building WWANs

Without a wired alternative, organizations frequently resort to WAN technologies to link together separate LANs. Contracting with a local telephone provider for a leased line presents a variety of drawbacks. Installation is typically expensive and rarely immediate. Monthly fees are often quite high for bandwidth that by LAN standards is very low. A wireless bridge can be purchased and then installed in less time and at a fraction of the cost. Once the initial hardware investment is made, there are no recurring charges, and today's wireless bridges provide the bandwidth one would expect from a technology rooted in data, rather than voice, communications.

In the same way that a commercial radio signal can be picked up in all sorts of weather miles from its transmitter, WLAN technology applies the power of radio waves to truly redefine the "local" in LAN. With a wireless bridge, networks located in buildings several miles from each other can be integrated into a single LAN (see Figure 2-4). When bridging between buildings with traditional copper or fiber optic cable, freeways, lakes, and even local governments can be impassible obstacles. A wireless bridge is a small device that acts as a wireless transmitter and receiver, and it transmits data through the air and requires no license or other permit.

Welcome to the Wireless Enterprise

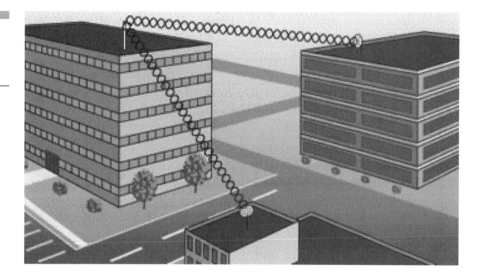

Figure 2-4
A building-to-building wireless network

WLAN technology is also used for bridging conventional wired networks together, which can be particularly useful if two sites have a line of sight with each other and the only wired bridging option involves monthly service charges to a Telco or other network provider. Many WLAN vendors offer bridging products based on the same spread spectrum technology as their WLANs. Using them is quite straightforward. First, you need to determine whether to use directional antennas, effective to about 100 meters, or directed antennas for a greater range. These antennas can easily cover distances of over a mile and with amplification (which is now allowed under recently relaxed FCC rules) can reach as far as 25 miles.

Specified throughputs range from 1 Mbps to 10 Mbps, depending on the product offering and price. The actual throughput will be less. In testing, 10 Mbps bridges show throughputs of just over 5 Mbps. After taking into account the range and throughput, consider any other bridging requirements you may have, such as support for routing and specific protocols. The cost of wireless connectivity at 5 Mbps is considerably less than the cost of wired connectivity at 5 Mbps.

To achieve building-to-building wireless connectivity, a wireless networking bridge equipped with a directional antenna provides the solution. Directional antennas require a line-of-sight connection, but can connect at distances of up to 1 mile away. Once the directional antennas are installed, all that is required is to connect them to the wireless access points and configure the settings. This enables WLAN connectivity in a campus environment with multiple buildings.

In-Building WLANs

WLAN technology can take the place of a traditional wired network or extend its reach and capabilities. Much like their wired counterparts, in-building WLAN equipment consists of PC Cards, a *personal computer interface* (PCI), and *Industry Standard Architecture* (ISA) client adapters, as well as access points[1], which perform functions similar to wired networking hubs or routers, depending on the features of the device. Similar to wired LANs for small or temporary installations, a WLAN can be arranged in a peer-to-peer or ad hoc topology using only client adapters. For greater functionality and range, access points can be incorporated to act as the center of a star topology and function as a bridge to an Ethernet network as well.

Within a building, wireless networking enables computing that is both mobile and connected. With a PC Card client adapter installed in a notebook or handheld PC, users can move freely within a facility while maintaining access to the network.

Applying WLAN technology to desktop systems provides an organization with flexibility impossible with a traditional LAN. Desktop client systems can be located in places where running cable is impractical or impossible. Desktop PCs can be redeployed anywhere within a facility as frequently as needed, making wireless ideal for temporary workgroups and fast-growing organizations (see Figure 2-5).

---

[1]Access points enable devices to connect to the network. They are covered in greater detail in later chapters.

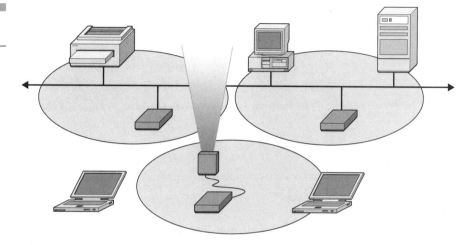

Figure 2-5
In-building WLAN

Cellular-based Wireless Technology

In the business world, connecting cell phones, pagers, and handhelds to the corporate network is not always a requirement. In most cases, the *total cost of ownership* (TCO) versus gains is the mitigating factor. Connecting other communication devices, such as pagers, cell phones, and other handhelds directly to the Internet, however, is useful.

Wireless Application Protocol (WAP)

WAP is a standard for providing cell phones, pagers, and other handheld devices with secure access to e-mail and text-based Web pages. Introduced in 1997 by Phone.com, Ericsson, Motorola, and Nokia, WAP provides a complete environment for wireless applications that includes a wireless counterpart of the *Transmission Control Protocol/Internet Protocol* (TCP/IP) and a framework for telephony integration such as call control and phone book access. WAP features the *Wireless Markup Language* (WML), which was derived from Phone.com's *Handheld Device Markup Language* (HDML) and is a streamlined version of *Hypertext Markup Language* (HTML) for small screen displays. It also uses WMLScript, a compact JavaScript-like language that runs in limited memory.

WAP also supports handheld input methods such as a keypad and voice recognition. Independent of the air interface, WAP runs over all the major wireless networks in place now. It is also device independent, requiring only a minimum functionality in the unit so that it can be used with myriad phones and handheld devices.

WAP applications run on the small screens of web-enabled cell phones. Web-enabled cell phones use the existing cellular infrastructure for data transmission and, as a result, use CDMA.

CDMA

CDMA is simply a technology for transmitting simultaneous signals over a shared portion of the spectrum, much the same way as FHSS and DSSS. The foremost application of CDMA is the digital cell phone technology from Qualcom that operates in the 800 MHz band and the 1.9 GHz *Personal Communication System* (PCS) band. CDMA phones are noted for their excellent call quality and long battery life.

CDMA is less costly to implement, requiring fewer cell sites than the *Global System for Mobile Communications* (GSM) and *Time Division Multiple Access* (TDMA) digital cell phone systems. CDMA also provides three to five times the calling capacity and provides more than 10 times the capacity of the analog cell phone system (the *Advanced Mobile Phone System* [AMPS]). CDMA has become widely used in North America and is also expected to become the *third-generation* (3G) technology for GSM.

Unlike GSM and TDMA, which divide the spectrum into different time slots, CDMA uses a spread spectrum technique to assign a code to each conversation. After the speech codec[2] converts voice to digital, CDMA spreads the voice stream over the full 1.25 MHz bandwidth of the CDMA channel, coding each stream separately so it can be decoded at the receiving end. The rate of the spreading signal is known as the *chip rate*, as each bit in the spreading signal is called a *chip* (no relation to an integrated circuit). All voice conversations use the full bandwidth at the same time. One bit from each conver-

---

[2] A code is a short form of a program that resides on a chip in the plane.

sation is multiplied into 128 coded bits by the spreading techniques, giving the receiving side an enormous amount of data it can average just to determine the value of one bit.

How CDMA Works   CDMA is a fascinating technology, and the example below shows you how calls from a base station are encoded and transmitted to a cell phone.

At the base station, each voice conversation is converted into digital code and compressed with a vocoder. The vocoder output is doubled by a convolutional encoder that adds redundancy for error checking. Each bit from the encoder is replicated 64 times and exclusive ORed with a Walsh code that is used to identify that call from the rest.

The output of the Walsh code is exclusive ORed with the next string of bits (*psuedo noise* [PN] sequence) from a pseudorandom noise generator, which is used to identify all the calls in a particular cell's sector. At this point, there are 128 times as many bits as there were from the vocoder's output. All the calls are combined and modulated onto a carrier frequency in the 800 MHz range.

At the receiving side, the received signals are quantized (turned into bits) and run through the Walsh code and PN sequence correlation receiver to recover the transmitted bits of the original signal. When 20 ms of voice data is received, a Viterbi decoder corrects the errors using the convolutional code, and it all goes to the vocoder, which turns the bits back into waveforms (sound).

CDMA transmission has been used by the military for secure phone calls. Unlike FDMA and TDMA methods, CDMA's wide spreading signal makes it difficult to detect and jam.

3G

The world is going wireless, and a new generation of mobile communications technologies is helping to change the way the world communicates. Called 3G, these technologies will enable the development of new wireless systems and devices that combine voice, Internet, and multimedia services on a device about the size of a cell phone (see Figure 2-6).

Figure 2-6
The analog to 3G evolution

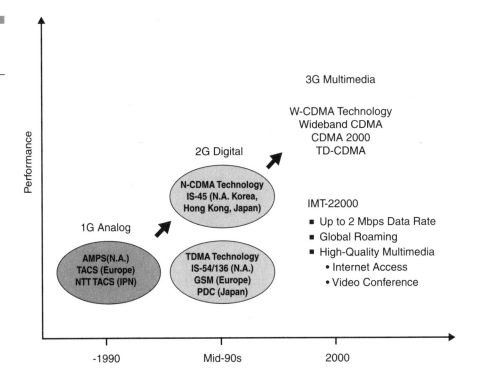

Three generations of mobile phones have emerged so far, each successive generation more reliable and flexible than the last (see Table 2-1):

- *Analog* cellular systems enabled you to make voice-only calls and typically only in any one country.
- *Digital* mobile phone systems added fax, data, and messaging capabilities, as well as voice telephone service in many countries.
- *3G* adds high-speed, broadband data transfer capabilities to mobile devices, enabling music, television, and the Internet to be accessed through a mobile terminal. With each new generation of technology, the services that can be deployed on them become more and more wide ranging and are limited only by one's imagination. We are reaching that stage with 3G.

During the first and second generations, different regions of the world pursued different mobile phone standards but are converging

Welcome to the Wireless Enterprise

Table 2-1

Wireless Evolution Summary

Generation	Technology Type	Era	Notes
1G	Analog	1980s	Designed to be used for voice, was easily susceptible to interference, and had multiple standards (NMT and TACS, for example).
2G	Digital	Early to mid-1990s	Designed to provide better quality voice connection, and was less susceptible to interference, but not immune. Multiple standards: GSM, CDMA, and TDMA.
2.5G	Digital/ higher data rate	Late 1990s	Introduced a new higher-speed data service designed to bridge the gap between 2G and 3G, enabling service providers time to update the infrastructure. 2.5G included such services as *General Packet Radio Service* (GPRS) and *Enhanced Data Rates for Global Evolution* (EDGE).
3G	Digital multimedia	Aprox. 2005	Voice- and data-centric with broadband data speeds for multimedia applications. Single, worldwide standard with multiple modes, enabling for the most flexibility.

to the common 3G standard for mobile multimedia based on CDMA technology. Europe pursued *Nordic Mobile Telephone* (NMT) and *Total Access Communications Systems* (TACS) for analog and GSM for digital, while North America pursued AMPS for analog and a mix of TDMA, CDMA, and GSM for digital. 3G will bring these incompatible standards together.

How 3G Works With 3G, the information is split into separate but related "packets" before being transmitted and reassembled at the receiving end. During packet switching, the packets are divided at the source and transported over the wireless network to the destination. During transportation of the data packets from the source to the end user, the packets get jumbled; that is, there is no guarantee that the packets will arrive in the same order in which they were sent. When the recipient receives all the packets, they are reassembled to form the original. All the packets are related and contain information to help them fit together, but the way they are transported and assembled varies.

Packet-switched data formats are much more common than their circuit-switched counterparts. Other examples of packet-based data standards include TCP/IP, X.25, Frame Relay, and *asynchronous transfer mode* (ATM). As such, although packet switching is new to the GSM world, it is well established elsewhere. In the mobile world, *Cellular Digital Packet Data* (CDPD), *Personal Digital Cellular Packet* (PDCP), GPRS, and wireless X.25 technologies have been in operation for several years. X.25 is the international public access packet radio data network standard.

Speeds of up to 2 Mbps are achievable with 3G. The data transmission rates will depend upon the environment the call is being made in; these types of data rates will only be available for indoors, stationary environments. For high mobility, data rates of 144 Kbps are expected to be available, which is only about three times the speed of today's fixed telecoms modems that have a transmission rate of 56 Kbps in ideal conditions.

3G facilitates several new applications that have not previously been readily available over mobile networks due to the limitations in data transmission speeds. These applications range from web browsing to file transfer to home automation or the ability to remotely access and control in-house appliances and machines. Because of the bandwidth increase, these applications will be even more easily available with 3G than they were previously with interim technologies such as GPRS.

Service Access   To use 3G, users specifically need the following:

- A mobile phone or terminal that supports 3G.
- A subscription to a mobile telephone network that supports 3G.
- Use of 3G must be enabled for that user. Automatic access to 3G may be allowed by some mobile network operators; others will charge a monthly subscription and require a specific opt-in to use the service as they do with other nonvoice mobile services.
- Knowledge of how to send and/or receive 3G information using his or her specific model of mobile phone, including software and hardware configurations (this creates a customer service requirement).
- A destination to send or receive information through 3G. From day one, 3G users can access any web page or other Internet applications, leading to an immediate critical mass of users.

These user requirements are not expected to change much for the meaningful use of 3G.

With the pending introduction of 3G technology, the world is expecting more sophisticated and diverse groups of products and services. 3G hardware and services have been tested in East Asia by major telecom players in mid-2001. Broadband-based services, such as video distribution, are scheduled to be tested around June 2002.

Broadband access to the Internet opens up a new range of services and enables the user a fast connection to the Internet. It extends the reach of the Internet but does not extend the reach of the corporate network. Accessing corporate data over a 3G device is limited to devices, not computers.

Summary

Different wireless networks have their advantages and disadvantages, that is different features and functionality. WLANs provide an excellent solution for wireless data access across a campus, office, or even home.

CHAPTER 3

Wireless LAN Primer

In the wired networking world, Ethernet has grown to become the predominant *local area network* (LAN) technology. Defined by the *Institute of Electrical and Electronics Engineers* (IEEE) as the 802.3 standard, Ethernet provides an evolved, high-speed, widely available, and interoperable networking standard. It has continued to evolve to keep pace with the data rate and throughput requirements of contemporary LANs. Originally providing 1 *megabit per second* (Mbps) transfer rates, the Ethernet standard evolved to include the 100 Mbps transfer rates required for network backbones and bandwidth-intensive applications. The IEEE 802.3 standard is an open architecture, decreasing barriers to market entry, and resulting in a wide range of suppliers, products, and price points from which Ethernet users can choose. Perhaps most importantly, conformance to the Ethernet standard allows for interoperability, enabling users to select individual products from multiple vendors while secure in the knowledge that they will all work together.

Since its inception as an industry standard, Ethernet has evolved. The next giant leap in evolution is a return to its origins. The result of the evolution is the 802.11 standard that provides network connectivity through a wireless physical link. Ironically, networking began its roots as a wireless technology, evolved into wired Ethernet, and today Ethernet and wireless networking have combined to become the *Wireless LAN* (WLAN) standard, 802.11.

The 802.11 standard promises to provide the advantages Ethernet networking with the capability to connect without wires. With 802.11, the two underlying layers, the physical media layer and data link layer, have been upgraded.

Physical Media Layer

There are three media that can be used for transmission over WLANs: *Infrared* (IR), *radio frequency* (RF), and microwave. In essence, all three media terms describe a chunk of the electromagnetic spectrum that is used for radio-based communication. The antenna and the RF or IR media serve as the new physical media and serve to replace the old, physical media wire. In 1985, the United

Wireless LAN Primer

States released the *industrial, scientific, and medical* (ISM) frequency bands. These bands are 902 to 928 MHz, 2.4 to 2.4853 GHz, and 5.725 to 5.85 GHz and do not require licensing by the *Federal Communications Commission* (FCC). This availability of a free and nonlicense-requiring band prompted most of the vendors to design WLAN products that operate within ISM bands. The FCC did put certain restrictions on the ISM bands, however.

In the United States, RF systems must implement spread spectrum technology (see Figure 3-1). RF systems must confine the emitted spectrum to a band. The RF energy is also limited to one watt of power. Microwave systems are considered very low power systems and must operate at 500 milliwatts or less. These restrictions do not impede the performance of wireless networks.

Radio Frequency (RF)

RF systems must use spread spectrum technology in the United States. This spread spectrum technology currently comes in two

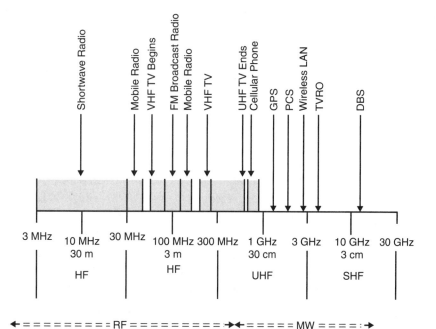

Figure 3-1
An electromagnetic spectrum

types: *direct sequence spread spectrum* (DSSS) and *frequency-hopping spread spectrum* (FHSS). There is a lot of overhead involved with spread spectrum, and so most of the DSSS and FHSS systems have historically had lower data rates than IR or microwave.

These RF systems are sometimes called *radio carriers* because they simply perform the function of carrying the data over using energy to a remote receiver. The data being transmitted is superimposed on the radio carrier so that it can be accurately extracted at the receiving end. This technique of imposing data on a carrier radio signal is generally referred to as modulation of the carrier by the information being transmitted. Once data is superimposed (modulated) onto the radio carrier, the radio signal occupies more than a single frequency because the frequency or bit rate of the modulating information adds to the carrier.

Multiple radio carriers can exist in the same space at the same time without interfering with each other if the radio waves are transmitted on different RFs. To extract data, a radio receiver tunes in (or selects) one RF while rejecting all other radio signals on different frequencies, much like a radio. In WLANs, these carriers are referred to as channels.

In a typical WLAN configuration, a transmitter/receiver (transceiver) device, called an *access point*, connects to the wired network from a fixed location using standard Ethernet cable. At a minimum, the access point receives, buffers, and transmits data between the WLAN and the wired network infrastructure. A single access point can support a small group of users and can function within a range of less than one hundred to several hundred feet, depending on the vendor. The access point (or the antenna attached to the access point) is usually mounted high but may be mounted essentially anywhere that is practical, as long as the desired radio coverage is obtained.

End users access the WLAN through WLAN adapters, which are implemented as PC cards in notebook computers, or use *Industry Structure Architecture* (ISA) or *Personal Component Interconnect* (PCI) adapters in desktop computers, or fully integrated devices within handheld computers. WLAN adapters provide an interface between the client *network operating system* (NOS) and the airwaves (via an antenna). The nature of the wireless connection is transparent to the operating system.

Narrowband Technology A narrowband radio system transmits and receives user information on a specific RF. Narrowband radio keeps the radio signal frequency as narrow as possible just to pass the information. Undesirable crosstalk between communications channels is avoided by carefully coordinating different users on different channel frequencies.

A telephone line, for example, is much like narrowband RF. When each home in a neighborhood has its own private telephone line, people in one home cannot listen to calls made to other homes. In a radio system, privacy and noninterference are accomplished by the use of separate RFs. The radio receiver filters out all radio signals except the ones on its designated frequency.

From a customer standpoint, a major drawback of narrowband technology is that the end user must obtain an FCC license for each site where it is deployed. This added burden of maintenance makes narrowband technology unpopular in WLANs.

Spread Spectrum Technology Most WLAN systems that use RFs or microwaves use spread spectrum technology, a wideband RF technique developed by the military for use in reliable, secure, mission-critical communications systems (see Figure 3-2). Spread spectrum is designed to trade off bandwidth efficiency for reliability, integrity, and security. More bandwidth is consumed than in the case of narrowband transmission, but the tradeoff produces a signal that is stronger and easier to detect, provided the receiver knows the parameters of the spread spectrum signal being broadcast. If a

Figure 3-2
Spread spectrum technology

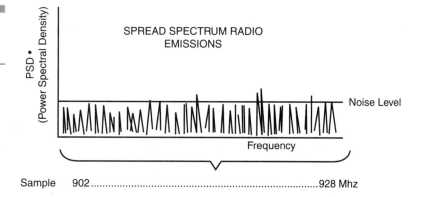

receiver is not tuned to the right frequency, a spread spectrum signal looks like background noise. Tuning the transmitter and receiver to the correct parameters is not the responsibility of the user; the devices negotiate the parameters themselves. There are two types of spread spectrum radio: frequency-hopping and direct sequence.

FHSS Technology FHSS uses a narrowband carrier that changes frequency in a pattern known to both transmitter and receiver. Properly synchronized, the net effect is to maintain a single logical channel (see Figure 3-3). To an unintended receiver, FHSS appears to be a short-duration impulse noise.

This technique splits the band into many small subchannels (1 MHz). The signal, then, hops from subchannel to subchannel, transmitting short bursts of data on each channel for a set period of time, called *dwell time*. The hopping sequence must be synchronized at the sender and the receiver, or information is lost. The FCC requires that the band is split into at least 75 subchannels and that the dwell time is no longer than 400 ms. Frequency-hopping is less susceptible to interference because the frequency is constantly shifting. This makes frequency-hopping systems extremely difficult to intercept. This feature gives frequency-hopping systems a high degree of security. In order to jam a frequency-hopping system with a radio jammer, the whole band must be jammed. These features are very

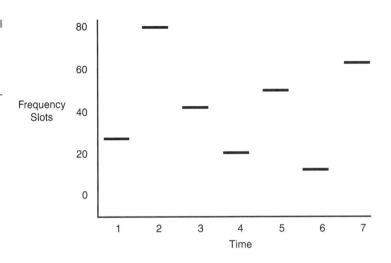

Figure 3-3
A frequency-hopping signal over time

attractive to agencies involved with law enforcement or the military. Many FHSS LANs can be colocated if an orthogonal hopping sequence is used. Because the subchannels are smaller than in DSSS, the number of colocated LANs can be greater with FHSS systems. Most new products in WLAN technology are currently being developed with FHSS technology.

DSSS Technology DSSS generates a redundant bit pattern for each bit to be transmitted. This bit pattern is called a *chip* (or chipping code). The longer the chip, the greater the probability that the original data can be recovered (and of course, the more bandwidth required). Even if one or more bits in the chip are damaged during transmission, statistical techniques embedded in the radio can recover the original data without the need for retransmission. To an unintended receiver, DSSS appears as low-power wideband noise and is rejected (ignored) by most narrowband receivers (see Figure 3-4).

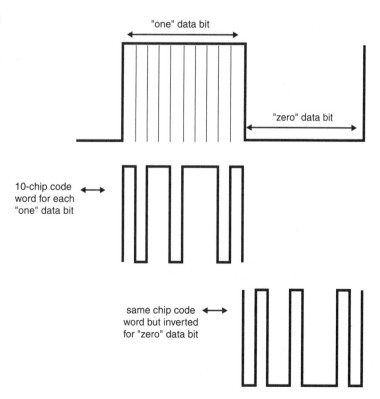

Figure 3-4
The DSSS

With DSSS, the transmission signal is spread over an enabled band (for example, 25 MHz). A random binary string is used to modulate the transmitted signal. This random string is called the *spreading code*. The data bits are mapped into a pattern of chips and mapped back into a bit at the destination. The number of chips that represent a bit is the *spreading ratio*. The higher the spreading ratio, the more the signal is resistant to interference. The lower the spreading ratio, the more bandwidth is available to the user. The FCC dictates that the spreading ratio must be more than 10. Most products have a spreading ratio of less than 20, and the new IEEE 802.11 standard requires a spreading ratio of 11. The transmitter and the receiver must be synchronized with the same spreading code. If orthogonal spreading codes are used, then more than one LAN can share the same band. However, because DSSS systems use wide subchannels, the number of colocated WLANs is limited by the size of those subchannels. Recovery is faster in DSSS systems because of the capability to spread the signal over a wider band.

IR Technology

A third technology, little used in commercial WLANs, is IR. IR systems use very high frequencies, just below visible light in the electromagnetic spectrum, to carry data. Like light, IR cannot penetrate opaque objects; it is either directed (line of sight) or diffuse technology. Inexpensive directed systems provide very limited range (3 feet) and typically are used for personal area networks but occasionally are used in specific WLAN applications. High performance directed IR is impractical for mobile users and is therefore used only to implement fixed subnetworks. Diffuse (or reflective) IR WLAN systems do not require line of sight, but cells are limited to individual rooms.

IR systems are simple in design and therefore inexpensive. They use the same signal frequencies used on fiber optic links. IR systems detect only the amplitude of the signal, and so interference is greatly reduced. These systems are not bandwidth limited and thus can achieve transmission speeds greater than other systems. IR transmission operates in the light spectrum and does not require a license from the FCC to operate, another attractive feature. There are two conventional ways to set up an IR LAN. The IR transmissions can be aimed, and this gives a good range of a couple of kilometers and can be used outdoors. It also offers the highest bandwidth and throughput. The other way is to transmit omnidirectionally and send the signal outward in a spherical pattern off of everything in every direction. This reduces coverage to 30 to 60 square feet. IR technology was initially very popular because it delivered high data rates at a relatively cheap price. The drawbacks to IR systems are that the transmission spectrum is shared with the sun and other things such as fluorescent lights. If there is enough interference from other sources, it can render the LAN useless. IR systems require an unobstructed line of sight because IR signals cannot penetrate opaque objects. This means that walls, cubicle dividers, curtains, and even fog can obstruct the signal.

It is not possible for the IR light to penetrate any solid material; it is even attenuated greatly by window glass, which is caused by the refraction of the light beams passing through the glass. It is really not considered to be a useful technology in comparison to RF for use in a WLAN system.

The application where IR comes into its element is as a docking function and in applications where the power available is extremely limited, such as a pager or PDA. There is a standard for such products called *IrDA* that has been championed by Hewlett Packard, IBM, and many others. This now is found in many notebook and laptop PCs and enables a connectionless docking facility at up to 1 Mbps to a desktop machine at up to 2 feet, line of sight.

Products based on IR technology are designed for point to point communications and not networks; this makes them very difficult to operate as a network but does offer increased security as only the user to whom the beam is directed can pick it up. Attempts to provide wider network capability by using a diffused IR system where the light is distributed in all directions have been developed and marketed, but they are limited to 30 to 50 feet and cannot go through any solid material. The main advantage of the point to point IR system, increased security, is undermined by the distributing of the light source as it can now be received by any body within range, not just the intended recipient.

Microwave

Microwave systems operate at less than 500 milliwatts of power in compliance with FCC regulations. Microwave systems are by far currently the fewest on the market and generally do not play a large part in 802.11. They use narrowband transmission with single frequency modulation and are set up mostly in the 5.8 GHz band. The big advantage to microwave systems is higher throughput achieved because they do not have the overhead involved with spread spectrum systems. Microwave systems will gain strength with the advent of 802.11a, the wireless networking standard allowing for speeds of up to 54 Mbps (see Figure 3-5).

Summary

WLANs come in many types: IR, microwave, and radio. Radio is further broken down into direct sequence and FHSS. The *Medium*

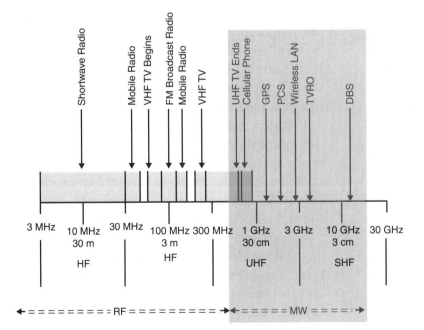

Figure 3-5
A microwave in the electromagnetic spectrum

Access Control (MAC) layer protocol used by WLANs as standardized in 802.11 is *Carrier Sense Multiple Access with Collision Avoidance* (CSMA/CA).

The Negroponte Switch Theory states that all things wired will be wireless, and all things wireless will become wired. This will certainly be true in the case of LANs. Traditional wired LANs will become a thing of the past as more and more users become mobile. LANs used to be defined by distance and spatial locality. Today, with the advances of wireless and virtual LAN technology, LANs are defined as a trust relationship regardless of location. Stationary users will become wireless once technology is able to increase throughput and data rates to levels that equal today's wired LANs.

Customers should be aware that WLAN systems from different vendors might not be interoperable for three reasons. First, different technologies, such as FHSS and DSSS, will not interoperate. A system based on FHSS technology will not communicate with another based on DSSS technology. Second, systems using different frequency bands will not interoperate even if they both employ the same technology, in wireless networking terms, an access point, and wireless *network interface cards* (NIC) will not be able to communicate if they are not on the same channel. Third, systems from different vendors may not interoperate even if they both employ the same technology and the same frequency band, due to differences in implementation by each vendor. This holds true only if the vendor(s) do not adhere to the 802.11 standard; all devices that are 802.11 compliant are interoperable.

MAC Protocol

Most wired LANs products use *Carrier Sense Multiple Access with Collision Detection* (CSMA/CD) as the MAC protocol. Carrier Sense means that the station will listen before it transmits. If there is already a station transmitting, then the station waits and tries again later. If no one is transmitting, then the station goes ahead and sends data. If two stations send at the same time, then the transmissions

Figure 3-6
The hidden node problem

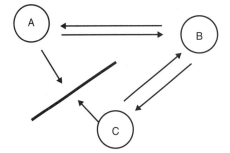

will collide, and the information will be lost; this is where Collision Detection comes in. The station will listen to ensure that its transmission made it to the destination without collisions. If a collision occurred, then the stations wait and try again later. The time the station waits is determined by the backoff algorithm. This technique works great for wired LANs, but wireless topologies can create a problem for CSMA/CD. The problem is the hidden node problem (see Figure 3-6).

In the hidden node problem, node C cannot hear node A. So if node A is transmitting, node C will not know and may transmit as well. This will result in collisions. The solution to this problem is CSMA/CA. CSMA/CA works as follows: The station listens before it sends. If someone is already transmitting, the station will wait for a random period and try again. If no one is transmitting, then it sends a short message, the *Ready to Send* message (RTS). This message contains the destination address and the duration of the transmission. Other stations now know that they must wait that long before they can transmit. The destination then sends a short message, which is the *Clear To Send* message (CTS). This message tells the source that it can send without fear of collisions. Each packet is acknowledged. If an acknowledgement is not received, the MAC layer retransmits the data. This entire sequence is called the *4-way handshake*, and this is the protocol that 802.11 chose for the standard handshaking (see Figure 3-7).

Wireless LAN Primer

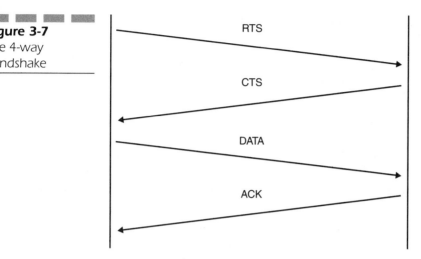

Figure 3-7
The 4-way handshake

Wireless Network Infrastructures

WLANs can be simple or complex. At its most basic, two PCs equipped with wireless adapter cards can set up an independent network whenever they are within range of one another. This is called a peer-to-peer network or ad hoc network. On-demand networks, such as in this example, require no administration or preconfiguration. In this case, each client would only have access to the resources of the other client and not to a central server or other resource on the corporate network.

Each computer, mobile, portable or fixed, is referred to as a *station* in 802.11. The difference between a portable and mobile station is that a portable station moves from point to point but is only used at a fixed point, such as a desktop computer with a WLAN card. Mobile stations can access the LAN during movement and move from access point to access point and include laptops that can move quickly without being shut down. When two or more stations come together to communicate with each other, they form a *Basic Service Set* (BSS). The minimum BSS consists of two stations, and 802.11 LANs use the BSS as the standard building block.

Ad Hoc Networking

A BSS that stands alone and is not connected to a base is called an *Independent Basic Service Set* (IBSS) or is referred to as an *ad hoc network*. An ad hoc network is a network where stations communicate only peer-to-peer and there is no base, such as an access point, and no one gives permission to talk (see Figure 3-8). Mostly, these networks are spontaneous and can be set up rapidly. Ad hoc or IBSS networks are characteristically limited both temporally and spatially.

One of the requirements of IEEE 802.11 is that it can be used with existing wired networks. 802.11 solved this challenge with the use of a portal. A *portal* is the logical integration between wired LANs and 802.11 and can serve as the access point to the *distribution system* (DS). All data going to an 802.11 LAN from an 802.X LAN must pass through a portal. It thus functions as bridge between wired and wireless, the same way a bridge for wired networks joins to wired networks. In most cases, no additional configuration is necessary to configure a bridge.

The implementation of the DS is not specified by 802.11. So a distribution system may be created from existing or new technologies. A point-to-point bridge connecting LANs in two separate buildings could become a DS. Although the implementation for the DS is not specified, 802.11 does specify the services that the DS must support.

Figure 3-8
Ad hoc networks

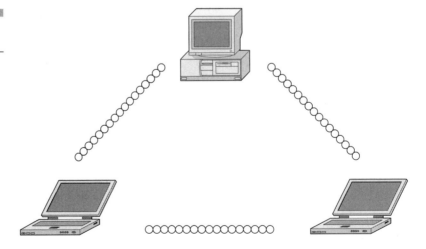

Services are divided into two sections, *Station Services* (SS) and *Distribution System Services* (DSS).

There are five services provided by the DSS: association, reassociation, disassociation, distribution, and integration; the first three services deal with station mobility. If a station is moving within its own BSS or is not moving, the station's mobility is termed *no-transition*. If a station moves between BSSs within the same *Extended Service Set* (ESS), its mobility is termed *BSS-transition*. If the station moves between BSSs of differing ESSs, it is *ESS-transition*. A station must affiliate itself with the BSS infrastructure if it wants to use the LAN. This is done by associating itself with an access point. Associations are dynamic in nature because stations move, turn on, or turn off. A station can only be associated with one at a given point in time. Multihoming a wireless station is not possible unless a separate wireless NIC is used for each access point. This ensures that the DS always knows the location of the station. Association supports no-transition mobility but is not enough to support BSS-transition. Reassociation enables the station to switch its association from one to another. Both association and reassociation are initiated by the station.

Disassociation is when the association between the station and the access point is terminated and can be initiated by either party, station, or access point. A disassociated station cannot send or receive data. ESS-transition is not supported, and a station can move to a new ESS but will have to reinitiate connections. Distribution and integration are the remaining DSSs. Distribution is simply getting the data from the sender to the intended receiver. The message is sent to the local access point (input access point) then distributed through the DS to the access point (output access point) that the recipient is associated with. If the sender and receiver are in the same BSS, the input and out access points are the same. So the distribution service is logically invoked whether the data is going through the DS or not. Integration occurs when the output access point is a portal. Thus, 802.x LANs are integrated into the 802.11 DS.

Station services are authentication, deauthentication, privacy, and *MAC Service Data Unit* (MSDU) Delivery. With a wireless system, the medium is not exactly bounded as with a wired system, and in order to control access to the network, stations must first establish their identity. This is much like trying to enter a radio net in the military. Before you are acknowledged and allowed to converse, you must first pass a series of tests to authenticate your identity. Authentication is simply the verification of the access point's identity. Once a station has been authenticated, it may then associate itself. The authentication relationship may be between two stations inside an IBSS or to the access point of the BSS. Authentication outside of the BSS does not take place. There are two types of authentication services offered by 802.11. The first is Open System Authentication, where any station that attempts to authenticate will receive authentication. The second type is Shared Key Authentication. In order to become authenticated, the users must be in possession of a shared secret key. The shared secret is implemented with the use of the *Wired Equivalent Privacy* (WEP) privacy algorithm. The shared secret is delivered to all stations ahead of time in some secure method (such as someone loading the secret onto each wireless network card). Deauthentication is when either the station or access point wants to terminate a station's authentication, and when this happens, the station is automatically disassociated. Privacy is an encryption algorithm that is used so that other 802.11 users cannot eavesdrop on LAN traffic. IEEE 802.11 specifies WEP as an optional algorithm to satisfy privacy. If WEP is not used, then stations are "in the clear" or "in the red," meaning that their traffic is not encrypted, and data transmitted in the clear is called *plaintext*. Data transmissions that are encrypted are called *ciphertext*. All stations start "in the red" until they are authenticated. MSDU delivery ensures that the information in the MAC service data unit is delivered between the medium access control service access points. To summarize, authentication is basically a network wide password. Privacy is whether or not encryption is used.

Most of the wireless mobile computing applications today require single hop wireless connectivity to the wired network. This is the traditional cellular network model that supports the current mobile computing needs by installing base stations and access points. In such networks, communication between two mobile stations completely relies on the wired backbone and the fixed base stations. However, sometimes no wired backbone infrastructure may be available for use by a group of mobile hosts. Also, there might be situations in which setting up of fixed access points is not a viable solution due to cost, convenience, and performance considerations. Still, the group of mobile users may need to communicate with each other and share information between them. In such situations, an ad hoc network can be formed. An ad hoc network is a temporary network, operating without the aid of any established infrastructure of centralized administration or standard support services regularly available on the *wide area network* (WAN) to which the hosts may normally be connected. Ad hoc networks are envisioned as infrastructureless networks where each node is a mobile router equipped with a wireless transceiver. A message transfer in an ad hoc network environment could either take place between two nodes that are within the transmission range of each other or between nodes that are indirectly connected via multiple hops through some intermediate nodes. The nodes that act as intermediate nodes in the data transfer process must be willing to participate in communication until successful message transfer has been accomplished. Applications of ad hoc networks include military tactical communication, emergency relief operations, commercial and educational use in remote areas, in meetings, and so on, where the networking is mission oriented and/or community based.

Ad hoc networks are comparable with the Internet, the radio Internet, fixed telephone, and cellular telephone network. The difference among them from the viewpoints of the service application is in the event that drives the information on each network. The telephone retrieves communication partners by the phone book. The phone book decides whom to communicate with, and the decision becomes an information-driving event. How to construct the phone book is the subject of the network service. The address book, which is equivalent to phone book, and the portal site, which is so called a *shared address book*, are used to decide where to send messages and from where to receive data. An ad hoc terminal itself searches other physically pluggable ad hoc terminals in the real world in the ad hoc network. Each station in the ad hoc network has the physical connection history of the way of the existence in the real world. The connection history drives information on the ad hoc network based with other terminal search information and also the connection condition at the present time. The development of the service history from the connection history is important to cause the behavior of the station inside the real world coordinated with service in the application on the upper class layer, as the temporary terminal sensing function and the connection condition control are important on the lower-class layer.

Infrastructure Networking

In infrastructure mode, clients talk to each other via a wireless access point, which has a standard cable connection to the wired network. Access points have a limited range to which their signal broadcasts. Installing more of these allows a wireless client to roam between microcells and communicate with the access point with the strongest signal.

When BSSs are interconnected, the network becomes the infrastructure. In 802.11, infrastructure networks have several elements. Two or more BSSs are interconnected using a DS. The concept of DS increases network coverage, and each BSS becomes a component of an extended, larger network. Entry to the DS is accomplished with the use of access points. An access point is an addressable station, much like a switch or router. Data moves between the BSS and the DS with the help of access points.

Creating large and complex networks using BSSs and DSs leads us to the next level of hierarchy, the ESS. The beauty of the ESS is that the entire network looks like an independent basic service set to the *Logical Link Control* (LLC) layer. This means that stations within the ESS can communicate or even move between BSSs transparently to the LLC.

Infrastructure networking is an 802.11 networking framework in which devices communicate with each other by going through an access point (see Figure 3-9). In infrastructure mode, wireless devices can communicate with each other or can communicate with a wired network through the access point. When one access point is connected to a wired network and one or more wireless stations, it is referred to as a BSS. An ESS is a set of two or more BSSs that forms a single subnetwork. Most corporate WLANs operate in infrastruc-

Wireless LAN Primer

Figure 3-9
An infrastructure network

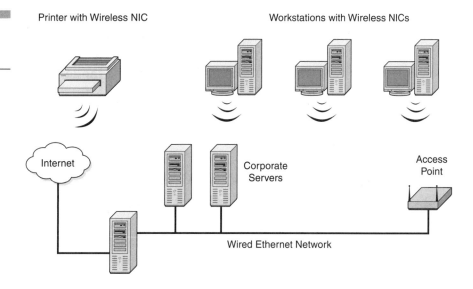

ture mode because they require access to the wired LAN in order to use services, such as file servers or printers, and enable access to the Internet.

An access point is a hardware device or a computer's software that acts as a communication hub for users of a wireless device to connect to a wired LAN. Access points are important for providing heightened wireless security and for extending the physical range of service of a wireless network. They are dedicated networking devices such as routers and switches. In the enterprise, a computer setup as a wireless access point is possible but not considered to be a best practice. For *Small Office/Home Office* (SOHO) users, server setup as an access point might have been a cost-effective alternative to separately purchasing an access point. However, the price point for most access points is at a level that makes them affordable without sacrificing reliability.

Selecting the Hardware

There are several vendors and manufacturers offering WLAN products that are 802.11 compliant. Offerings include wireless access points, workgroup bridges for connectivity to the Ethernet network, and client adapters for the computing devices that require wireless network connectivity. For best practices, selecting wireless devices that are 802.11 compliant and using the same transmission technology, FHSS or DSSS, and the same media, RF or IR, will ensure ease of deployment and compatibility.

Choosing the Best Configuration for the Enterprise

For the enterprise, a well-designed infrastructure network is the preferred solution. Designed properly, implementing a wireless infrastructure network allows all users access to the corporate network and its resources. There are two basic network infrastructures, ad hoc networking and infrastructure networking. For the enterprise, an infrastructure network provides the best solution of security, scalability, and availability.

Coverage Options

The network architecture options of wireless stations and access points provide for a variety of coverage alternatives and flexibility. The network can be designed to provide a wide coverage area with minimal overlap or a narrow coverage area with heavy overlap. A narrow coverage area with heavy overlap improves network performance and protection against downtime if a component fails.

Minimal Overlap Coverage Option

By arranging the access points so that the overlap in a coverage area is minimized, a large area can be covered with minimal cost (see Figure 3-10). The total bandwidth available to each wireless client

Wireless LAN Primer

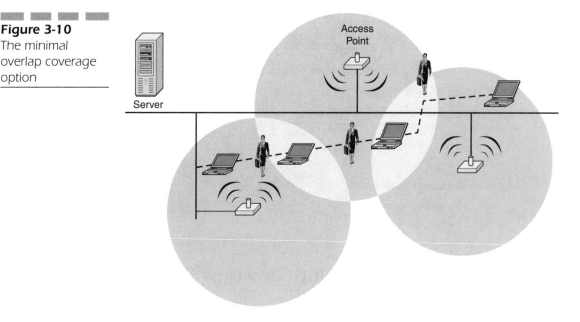

Figure 3-10
The minimal overlap coverage option

device depends on the amount of data each mobile station needs to transfer and the number of stations located in each cell. Seamless roaming is supported as a client device moves in and out of range of each access point, thereby maintaining a constant connection to the wired LAN. Each device in the radio network must be configured with the same *service set identifiers* (SSIDs) to provide roaming capability.

Multiple Overlapping Networks Coverage Option

Multiple networks can operate in the same vicinity (see Figure 3-11). The architecture provides multiple channels that can exist in the same area with virtually no interference to each other. In this mode, each system must be configured with different SSIDs and different channels, which may (depending on configurations) prevent clients from roaming to access points of a different wireless network.

Figure 3-11
The multiple overlapping networks

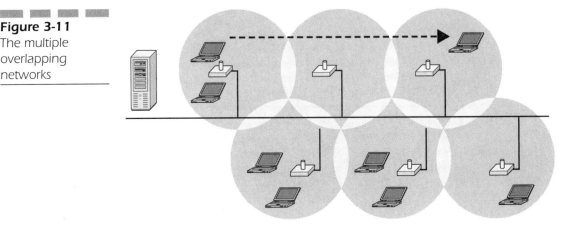

Heavy Overlap Coverage Option

By arranging the access points so the overlap in the coverage area is nearly maximized, a large number of mobile stations can be supported in the same wireless infrastructure. However, devices in overlapping coverage areas on the same frequency will detect adjacent cell traffic and delay transmissions that would cause collisions. This configuration reduces the radio system throughput, and heavy cell overlap is not recommended for maximum system throughput (see Figure 3-12).

Because of the redundancy in coverage overlap, network access is not lost if an access point fails. Upon failure of the access point, the station automatically roams to an operational access point. With this architecture, each device in the RF network must be configured with the same SSID to provide the roaming capability.

Site Surveys

Because of differences in component configuration, placement, and physical environment, every network application is a unique installation. Before installing multiple access points, perform a site survey to determine the optimum utilization of networking components and to maximize range, coverage, and network performance.

Wireless LAN Primer

Figure 3-12
The heavy overlap coverage option

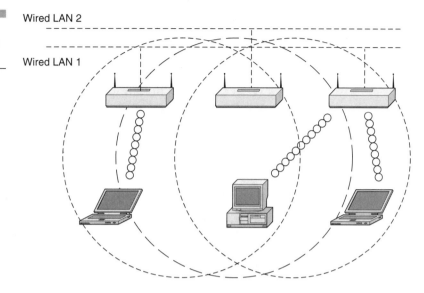

Consider the following operating and environmental conditions when performing a site survey:

- ***Data rates*** Sensitivity and range are inversely proportional to data bit rates. The maximum radio range is achieved at the lowest workable data rate. A decrease in receiver threshold sensitivity occurs as the radio data increases.

- ***Antenna type and placement*** Proper antenna configuration is a critical factor in maximizing radio range. As a general rule, range increases in proportion to antenna height.

- ***Physical environment*** Clear or open areas provide better radio range than closed or filled areas. Also, the less cluttered the work environment, the greater the range.

- ***Obstructions*** A physical obstruction, such as metal shelving or a steel pillar, can hinder performance of wireless devices. Avoid locating the devices in a location where there is a metal barrier between the sending and receiving antennas.

- ***Building materials*** Radio penetration is greatly influenced by the building material used in construction. For example, drywall construction enables greater range than concrete blocks. Metal or steel construction is a barrier to radio signals.

- ***Power for the access point*** Like most networking devices, access points require power as well as a network connection. There are some models of access points and switches that support through the Ethernet cable. If installing an access point with a switch that does not support this, a power connection, the standard 110V AC in the United States must be supplied to the location of the access point.

Range and Coverage in the Real World

The distance that RF and IR signals can travel is a function of product design—including transmitted power, receiver design, and antennae design—and obstacles in the propagation path, especially in indoor environments. Interactions with typical building objects, including walls, metal, and even people, can affect how the signal propagates and thus what range and coverage a particular system achieves. Solid objects block an IR signal, which imposes additional limitations and makes the technology difficult to use in the real world. Devices can communicate only if they have line-of-site, that is, the path that the IR wave must travel is unobstructed. Most WLAN systems use RF because radio waves can penetrate most indoor walls and obstacles. The range (or radius of coverage) for typical WLAN access point varies from under 100 feet to a little more than 300 feet. However, as the range increases, the speed of the connection degrades. For optimal performance, coverage can be extended, and true freedom of mobility via roaming is provided through microcells.

Microcells and Roaming

WLAN communication is limited by how far signals carry for given power output. WLANs use coverage cells, (microcells) created by access points, similar to the cellular telephone system, to extend the range of wireless connectivity. At any point in time, a mobile PC

Wireless LAN Primer

equipped with a wireless network adapter is associated with a single access point and its microcell or an area of coverage. Individual microcells overlap to enable continuous communication within wired network. They handle low-power signals and hand off users as they roam through a given geographic area.

Multiple microcells are created by using multiple access points to create several coverage areas, sometimes overlapping, that allow for coverage of the enterprise (see Figure 3-13). Creating overlap provides an added benefit; the wireless network card will automatically select the access point that provides the best service, speed, and reliability. This ensures the coverage area has no gaps in coverage and a level of redundancy.

A typical mixed wired/wireless LAN configuration is analogous to a standard cellular phone system. The Ethernet-wired LAN functions as a backbone trunking line interconnecting all access points (the equivalent of cellular base stations in a cellular phone system). Transparent roaming across different WLAN cells is supported as part of the IEEE 802.11 standard, much the same way as driving through multiple cellular coverage cells allows for seamless roaming.

In many WLAN applications with multiple access points, users should be able to maintain a continuous connection while roaming

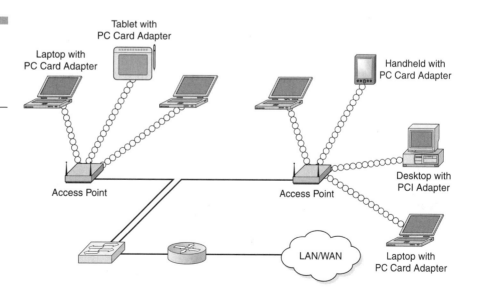

Figure 3-13
An infrastructure network with overlap and roaming

Figure 3-14
Roaming in a wireless network

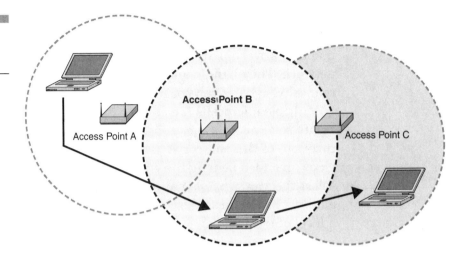

from one physical area to another (see Figure 3-14). In doing so, the station may well move from the coverage of one access point to another. Nearly all WLAN manufacturers support this kind of roaming through a process by which the mobile nodes automatically register with the new access point. What administrator and network engineers will need to consider in network planning is how the infrastructure network is divided into subnets. If one access point is on one subnet and another access point is on another subnet, traffic will have to cross a router, something that most WLAN vendors currently do not support. The two possible workaround to this limitation are

- Connect all access points to one subnet, which is a straightforward solution. All access points need to be connected to one backbone or switch, and the proper subnet mask needs to be entered in the access point's configuration.
- Use mobile IP if your network protocol is IP.

Mobile IP

If your roaming wireless users sit down, turn on their portable computers, establish a wireless networking session, conduct their business, log off, and then turn off their machines, mobility may not be an issue, especially if the users obtain IP addresses dynamically through the *Dynamic Host Control Protocol* (DHCP). However, if IP network users keep their machines on, perhaps use handheld computers and wander from one area to another, they may ultimately connect with an access point that is connected to a different subnet. Because IP addresses by design refer to a particular subnet, IP traffic will not be able to find its way to the new location. Mobile IP enables a workaround to this limitation. A recently completed Internet standard, RFC 2002, Mobile IP provides a mechanism by which hosts belong to a home network. When the host roams to a new network, it registers with its home agent. The home agent then intercepts traffic sent to the mobile node, encapsulates it in another IP packet, and forwards it to a foreign agent, a special node (installed in a workstation, router, or access point) that forwards the packets to the mobile node. Traffic originated by the mobile node can travel back directly and does not have to be sent via the home node, resulting in triangular routing of traffic.

How It Works

Mobile IP, an extension to IP, is a recent Internet standard specified in RFC 2002, "IP Mobility Support." The concept behind Mobile IP is simple, though a number of complications may arise when implementing it.

Mobile IP consists of three components:

- Mobile node
- Home agent
- Foreign agent

The mobile node is built into a TCP/IP stack or can exist as a "shim" under a TCP/IP stack. The home agent operates on a router or a workstation on the mobile node's home subnet. The foreign agent operates on a router or workstation on a foreign network where the mobile node is visiting or on the mobile node itself under certain conditions. These are the only elements required to implement a Mobile IP solution. No other changes are needed in any other part of the network, including routers or other systems, such as DNS.

When a mobile station comes up on a network, it first determines whether it is on the home network or on a foreign network. It accomplishes this by listening for a local broadcast message from a home agent or foreign agent, or alternatively, it can solicit an agent advertisement message. These initial and subsequent registration messages are based on extensions to *Internet Control Message Protocol* (ICMP) Router Discovery specified by RFC 1256.

When the mobile station is on its home subnet—the one specified by its IP address—the mobile node informs the home agent of its presence. From there, IP addressing and datagram delivery work as they would without Mobile IP. The situation changes when the mobile node connects to a foreign network. A foreign network being one that does not have the same IP address as the mobile station. There it obtains a "care of address," which is the foreign agent's IP address. The mobile node registers with its home agent and gives the home agent its care of address. Alternatively, if DHCP is available on the foreign network, the mobile node can obtain a temporary address, register this with the home agent, and act as its own foreign agent.

Once the mobile station has registered with the home agent, IP traffic addressed to the mobile station is received by the home agent, encapsulated in another IP datagram, and then tunneled to the foreign agent. The foreign agent forwards the datagram to the mobile station. Two forms of encapsulation are specified in related standards RFC 2003 (IP Encapsulation within IP) and 2004 (Minimal Encapsulation within IP).

For sending data traffic in the reverse direction, the mobile station can bypass the home agent and send datagrams directly to their destination. This results in a triangular routing of traffic, which is not necessarily efficient, but it is effective. In addition, when a mobile station changes its location, it can register with a new foreign agent, though traffic directed by the home agent to the old foreign agent will be lost until the new mobile node has registered its location.

Although Mobile IP is not the most efficient solution, it is effective in enabling users the ability to roam from access point to access point. Allowing users to move throughout the enterprise with no changes in configuration is required (see Figure 3-15).

Wireless LAN Primer

Figure 3-15
The mobile IP Summary

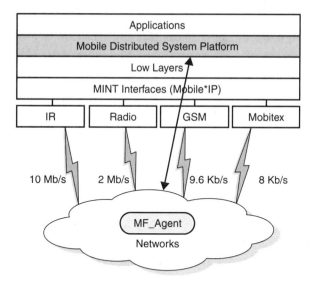

Security

A secure wireless network is achieved by addressing the chief concern of access control and privacy. Access control ensures that sensitive data can be accessed only by authorized users. Privacy ensures that transmitted data can be received and understood only by the intended audience.

Access to a wired LAN is governed by access to a specific Ethernet port for that LAN. Therefore, access control for a wired LAN often is viewed in terms of physical access to LAN ports. Similarly, because data transmitted on a wired LAN is directed to a particular destination, privacy cannot be compromised unless someone uses specialized equipment to intercept transmissions on his way to his destination. In short, a security breach on a wired LAN is possible if the LAN is physically compromised, or it is connected to an unsecured network segment. It is possible for a security breach to come from the Internet if the network is connected to the Internet and not secured properly at that connection.

With a WLAN, transmitted data is broadcast over the air using radio waves, so it can be received by any WLAN client in the area served by the data transmitter. Because radio waves travel through ceilings, floors, and walls, transmitted data may reach unintended recipients on different floors and even outside the building of the transmitter. Similarly, data privacy is a genuine concern with WLANs because there is no way to direct a WLAN transmission to only one recipient.

The IEEE 802.11b standard includes components for ensuring access control and privacy, but to ensure the highest level of security, these components must be deployed on every device in a WLAN. An organization with hundreds or thousands of WLAN users needs a solid security solution that can be managed effectively from a central point of control. Some cite the lack of centralized security as the primary reason why WLAN deployments have been limited to relatively small workgroups and specialized applications.

First-generation WLAN Security

The IEEE 802.11b standard defines two mechanisms for providing access control and privacy on WLANs: SSIDs and WEP. Another mechanism to ensure privacy through encryption is to use a *virtual private network* (VPN) that runs transparently over a WLAN. The use of a VPN is independent of any native WLAN security scheme, and it's the same for WLAN and wired LAN.

SSID

One commonly used WLAN feature is a naming handle called a SSID, which provides a basic level of access control. A SSID is a common network name for the devices in a WLAN subsystem; it serves to logically segment that subsystem. The use of the SSID as a handle to permit/deny access is dangerous because the SSID typically is not well secured. An access point, the device that links wireless clients to the wired LAN, usually is set to broadcast its SSID in all of its beacons.

Wireless LAN Primer

The latest release of Microsoft Windows, XP, discovers the network SSID automatically and makes connecting to an unsecured WLAN effortless. This solution is ideal for public access places, such as airports and hotels, but not for a wireless enterprise.

WEP

The IEEE 802.11b standard stipulates an optional encryption scheme called WEP, which offers a mechanism for securing WLAN data streams. WEP uses a symmetric scheme where the same key and algorithm are used for both encryption and decryption of data. The goals of WEP include

- ***Access control*** Prevent unauthorized users, who lack a correct WEP key, from gaining access to the network.
- ***Privacy*** Protect WLAN data streams by encrypting them and allowing decryption only by users with the correct WEP keys.

Although implementing WEP on a network is optional, support for WEP with 40-bit encryption keys is a requirement for Wi-Fi certification by WECA, so WECA members invariably support WEP. Some vendors implement the computationally intense activities of encryption and decryption in software, while others, like Cisco Systems, use hardware accelerators to minimize the performance degradation of encrypting and decrypting data streams.

The IEEE 802.11 standard provides two schemes for defining the WEP keys to be used on a WLAN. With the first scheme, a set of as many as four default keys are shared by all stations—clients and access points—in a wireless subsystem. When a client obtains the default keys, that client can communicate securely with all other stations in the subsystem. The problem with default keys is when they become widely distributed, they are more likely to be compromised. In the second scheme, each client establishes a key mapping relationship with another station. This is a more secure form of operation because fewer stations have the keys, but distributing such unicast keys becomes more difficult as the number of stations increases.

WEP is used to protect authorized stations from eavesdroppers. WEP is reasonably strong, but like all things, can be broken in time. The relationship between breaking the algorithm is directly related to the length of time that a key is in use. So WEP allows for changing of the key to prevent brute force attack of the algorithm. WEP can be implemented in hardware or in software. One reason that WEP is optional is because encryption may not be exported from the United States. This enables 802.11 to be a standard outside the United States without the WEP encryption.

Authentication

A client cannot participate in a WLAN until that client is authenticated, which is the first step in joining a wireless network. The IEEE 802.11b standard defines two types of authentication methods: open and shared key. The authentication method must be set on each client, and the setting should match that of the access point with which the client wants to associate.

With open authentication, which is the default, the entire authentication process is done in clear-text, and a client can associate with an access point even without supplying the correct WEP key. With shared-key authentication, the access point sends the client a challenge text packet that the client must encrypt with the correct WEP key and return to the access point. If the client has the wrong key or no key, it will fail authentication and will not be allowed to associate with the access point.

Some WLAN vendors support authentication based on the physical address, or MAC address, of a client. An access point will enable association by a client only if that client's MAC address matches an address in an authentication table used by the access point.

A Complete Security Solution

What is needed is a WLAN security solution that uses a standards-based and open architecture to take full advantage of 802.11b security elements, provide the strongest level of security available, and

Wireless LAN Primer

ensure effective security management from a central point of control. A promising security solution implements key elements of a proposal jointly submitted to the IEEE by Cisco Systems, Microsoft, and other organizations (see Figure 3-16). Central to this proposal are the following elements:

- *Extensible Authentication Protocol* (EAP), an extension to *Remote Access Dial-In User Service* (RADIUS) that can enable wireless client adapters to communicate with RADIUS servers
- IEEE 802.1X, a proposed standard for controlled port access

When the security solution is in place, a wireless client that associates with an access point cannot gain access to the network until the user performs a network logon. When the user enters a username and password into a network logon dialog box or its equivalent, the client and a RADIUS server (or other authentication server) perform a mutual authentication, with the client authenticated by the supplied username and password. The RADIUS server and client then derive a client-specific WEP key to be used by the client for the current logon session. All sensitive information, such as the password, is protected from passive monitoring and other methods of attack. Nothing is transmitted over the air in the clear.

The sequence of events follows:

1. A wireless client powers up and associates with an access point.
2. The access point blocks all attempts by the client to gain access to network resources until the client logs onto the network.
3. The user on the client supplies a username and password in a network logon dialog box or its equivalent.
4. Using 802.1X and EAP, the wireless client and a RADIUS server on the wired LAN perform a mutual authentication through the access point. One of several authentication methods or types can be used. With the Cisco authentication type, the RADIUS server sends an authentication challenge to the client. The client uses a one-way hash of the user-supplied password to fashion a response to the challenge and sends that response to the RADIUS server. Using information from its user database, the RADIUS server creates its own response and compares that to

the response from the client. Once the RADIUS server authenticates the client, the process repeats in reverse, enabling the client to authenticate the RADIUS server.

5. When mutual authentication is successfully completed, the RADIUS server and the client determine a WEP key that is distinct to the client and provides the client with the appropriate level of network access, thereby approximating the level of security inherent in a wired switched segment to the individual desktop. The client loads this key and prepares to use it for the logon session.

6. The RADIUS server sends the WEP key, called a session key, over the wired LAN to the access point.

7. The access point encrypts its broadcast key with the session key and sends the encrypted key to the client, which uses the session key to decrypt it.

8. The client and access point activate WEP and use the session and broadcast WEP keys for all communications during the remainder of the session (see Figure 3-17).

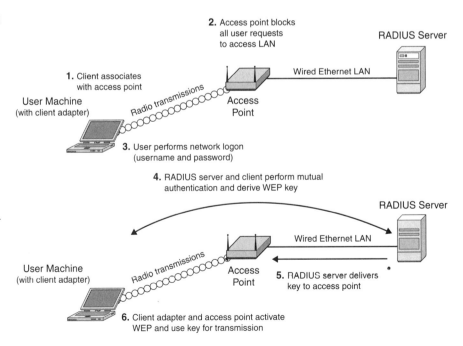

Figure 3-16
Cisco security solution, association, and authentication are based on username and password, and each user gets a unique, session-based encryption key.

Wireless LAN Primer

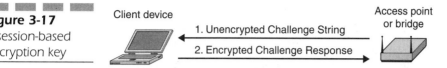

Figure 3-17
A session-based encryption key

Support for EAP and 802.1X delivers on the promise of WEP, providing a centrally managed, standards-based, and open approach that addresses the limitations of standard 802.11 security. In addition, the EAP framework is extensible to wired networks, enabling an enterprise to use a single security architecture for every access method.

It is likely that dozens of vendors will implement support for 802.1X and EAP in their WLAN products once the standard is approved.

With the 802.1X WLAN security solution in place, an organization

- Minimizes the security threats of lost or stolen hardware, rogue access points, and hacker attacks
- Uses user-specific, session-based WEP keys created dynamically at user logon, not static WEP keys stored on client devices and access points
- Manages the security for all wireless users from a central point of control

The proposed wireless security services closely parallel security available in a wired LAN, fulfilling the need for a consistent, reliable, and secure mobile networking solution.

Integration with Existing Networks

Although a wireless network can operate as a standalone network, chances are it will not be very useful without the capability to connect it to the wired infrastructure. Connecting the wireless network to the wired network will extend the reach of the existing wired network and can be easily accomplished by using a wireless access point. An access point is a wireless networking device that can serve as a bridge for wireless traffic from the wireless network to the wired network. The WLAN then appears as one network segment in the overall network. Purchasing the access point from the same manufacturer as the wireless NIC, with standards such as IEEE 802.11 and industry interoperability, initiatives such as the WLAN interoperability forum are no longer a requirement. Most access points bridge WLANs into Ethernet networks. Access points also exist that enable bridging a wireless network with a wired Token Ring network.

Number of Clients per Access Point

Each access point can accommodate several clients; the specific number depends on the amount of data (bandwidth) being used. When multiple clients connect to an access point, the 11 Mbps bandwidth must be shared among all connecting devices. Streaming multimedia applications places high bandwidth demands on the wireless network. In the real world, a single access point can serve from 15 to 50 client devices, with no loss in apparent connection speed to the user.

Throughput

As with wired LAN systems, actual throughput in WLANs is product, configuration, and environment dependent. Factors that affect throughput include the number of users, propagation factors such as range and multipath, the type of WLAN system used, as well as the latency and bottlenecks on the wired portions of the LAN. Data rates for the most widespread commercial WLANs are in the 5 Mbps range. Users of traditional Ethernet or Token Ring LANs generally experience little difference in performance when using a WLAN. WLANs provide throughput sufficient for the most common LAN-based office applications, including email exchange, access to shared peripherals, Internet access, and access to multiuser databases and applications.

Wireless LAN Primer

As a point of comparison, it is worth noting that state-of-the-art V.90 modems transmit and receive at optimal data rates of 56.6 Kbps. In terms of throughput, a WLAN operating at 1.6 Mbps is almost 30 times faster.

Getting Killer Throughput

A wireless network card or access point's specification data rates are specified as a function of distance. Data rate drops off very quickly with distance for 802.11b products, and the change is somewhat abrupt or stepped (see Figure 3-18).

This chart from Compaq's WL Series Reviewer's Guide gives you some idea of what to expect in a majority of 802.11b products, but like all manufacturer specs, the actual effective data rate will be different in the real world.

For coverage area planning in the real world, it is best to count on achieving about 50 to 75 percent of the published specification. This allows for coverage overlap and provides an excellent estimate of what real-world performance may be like. Once data rate and distance is plotted, testing the connection speed at various distances will help verify the estimates. Although many of the tips refer to an access point, they all apply equally to a wireless router, gateway, or bridge.

Figure 3-18
The range and coverage

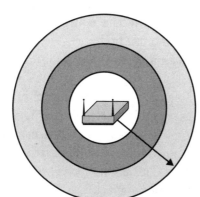

Range	Rate
Up to 100°	11 Mb/s
■ Up to 150°	5.5 Mb/s
□ Up to 300°	2 Mb/s

The best place to put your access point is as close to the center of the area that you want to cover and as high as possible. The preferred location is above the ceiling tiles, in the center of the coverage area.

The worst place for high throughput (weakest signal) is directly under an access point, assuming that the access point's antenna is vertically oriented and omnidirectional.

A best practice would be to orient the access point or wireless router's antenna(s) vertically. Keep antennas away from large metal objects like filing cabinets, from areas with operating microwave ovens or 2.4 GHz cordless phones, and from large containers of water such as fish tanks or water heaters, for example. Dealing with wireless networking equipment requires the same precautions as dealing with other wireless equipment like cordless phones and radios. If an access point is purchased that has a built in antenna, like the Buffalo Technology Air Station, mount the access point as recommended by the manufacturer. This will provide the best orientation for the antenna and the best coverage.

Most PC cards use an integrated antenna that is fairly directional in the horizontal plane. Unfortunately, the horizontal orientation of these PC card antennas is not optimal. Wireless network cards for laptops (PC Cards) work well but may provide a coverage area less than what the specifications dictate. Vertically oriented antennas in network cards would work better. At this time, PC cards with vertically oriented antennas are not being produced by a majority of manufacturers.

Avoid antenna placement close to an outside wall. This will result in greater security and efficiency. Also, if there is a need for users to connect while outdoors, an access point placed near a window will enable the signal to pass outside.

Several manufactures also provide a signal booster or extended range antennas. Buffalo Technology, ORiNOCO, and Cisco all provide solutions for upgrading the antenna for increased range of the WLAN. Although it is possible to tweak the performance of wireless networks by improving the antenna of the access point and the client (laptop, desktop, or other device), the best practice is to improve the antenna access point as much as possible. Keeping the original antenna on the wireless network card will ease the time of administration and updating required.

Integrity and Reliability of the Wireless Enterprise

Wireless data technologies have been proven through more than 50 years of wireless application in both commercial and military systems. Although radio interference can cause degradation in throughput, such interference is rare in the workplace and with 802.11 products. Robust designs of proven WLAN technology and the limited distance over which signals travel result in connections that are far more robust than cellular phone connections and provide data integrity performance equal to wired networking.

With the 802.11b implementation, the wireless infrastructure is transparent to the user and to the operating systems. Wireless networking systems are designed to be substituted for the wired Ethernet infrastructure that sits on Layer 1 and Layer 2 of the OSI model. This enables the operating system and applications to work without the need for special drivers specific to wireless networking and without the devices having to know about the wireless transport layers.

Summary

In summary, to address the security concerns raised in this section, a WLAN security scheme should

- Base WLAN authentication on device-independent items, such as usernames and passwords, which users possess and use regardless of the clients on which they operate
- Support mutual authentication between a client and an authentication (RADIUS) server
- Use WEP keys that are generated dynamically upon user authentication, not static keys that are physically associated with a client
- Support session-based WEP keys

First-generation WLAN security, which relies on static WEP keys for access control and privacy, cannot address these requirements.

- Data transmission rates decrease as users get farther from the access point. Refer to Table 3-1 for a summary of the speeds and most probable common speeds.

- It is important to note that actual data transmission rates may vary considerably, depending on number of users, type of usage (that is, file sizes and frequency of down/up loading), speed of network, and so on. On *all* networks (wired and wireless), some of the data being transmitted is "overhead," which controls and manages data flow. This overhead means that file transfer rate (for example, downloading a file from a shared drive) will not reach 11 Mbps. These issues are accepted and understood by IT professionals and network managers.

Table 3-1 Summary of Speeds

Speed Options	Range in Meters (Feet) Open Office	Range in Meters (Feet) Semi-Open Office	Closed Office	Receiver Sensitivity
11 Mbps	160 m (525 ft)	50 m (165 ft)	25 m (80 ft)	-82
5.5 Mbps	270 m (885 ft)	70 m (230 ft)	35 m (115 ft)	-87
2 Mbps	400 m (1300 ft)	90 m (300 ft)	40 m (130 ft)	-91
1 Mbps	550 m (1750 ft)	115 m (375 ft)	50 m (165 ft)	-94

CHAPTER 4

Access Point Installation

Deploying a *wireless local area network* (WLAN) is relatively straightforward. For wireless computers, you must install the wireless network card, as well as the appropriate software drivers that your network protocol stacks use to access the wireless card. These drivers also come with the card. The installation and configuration of a wireless network card is no different than that of a wired LAN card. In simplest terms, just install the hardware, appropriate drivers, and management software. There are two wireless network architectures, and hoc and infrastructure.

Ad hoc networks work well for teams of people that need impromptu data-sharing and collaboration capabilities, but they do not provide the robustness, security, and features of an infrastructure network. For an enterprise, an infrastructure architecture offers many benefits.

When deploying an infrastructure-based network, the challenge comes in knowing where to install the access points for the wireless computers and securing the wireless network. A well-thought-out deployment of the access points will ensure that a network provides users with the best possible coverage.

The first step is to define exactly where you want coverage, which requires a site survey. It is important to place the access points so that the area in which the wireless computers will operate is covered. Wireless signals can penetrate walls, sometimes several walls, but how well they do so depends on how the building was constructed and the materials used in the construction. To verify that you have coverage for the area you want, different vendors provide different tools, but all vendors provide software that allows testing of the wireless connection.

One approach is to use two portable computers with wireless network cards operating in an ad hoc network that transmit and receive test data. In an ad hoc network, the computers operate much the same way as a wireless access point. Using diagnostic software provided by the vendor, you should determine a coverage area for a potential access point by keeping one portable computer fixed and moving around with the other. Most software comes with a software-based signal strength meter that shows the strength of the signal and, hence, the connection speed. Check the documentation to see which tools they provide and what approach they recommend for deploying their access points.

Access Point Installation and Configuration Outline

The access point will come in a box with the software CD and documentation. On the CD will be more information, software drivers, and additional documentation. In addition to the access point, power connector, and CD, an access point installation also requires the following:

- A computer connected to the same wireless or wired network as the access point
- Category 5, straight-through Ethernet cable for connecting the access point to the network and a power outlet
- The *Medium Access Control* (MAC) address from the label on the bottom of the access point (such as 07596674654b)
- The following information from your network system administrator:
    - The case-sensitive wireless *service set identifier* (SSID) for your radio network
    - A unique IP address for the access point (such as 192.168.1.88) if there is no *Dynamic Host Configuration Protocol* (DHCP) server connected to the network
    - A default gateway and subnet mask if the access point is not on the same subnet as your PC

Plan to configure the access point before mounting it on a ceiling or wall. Some steps, requiring physical access to the access point, are easier to perform if the access point is accessible.

Connecting and Powering Up

The first step is to connect the access point to the wired network and power up.

1. Remove the access point, power connection, software CD, and Ethernet cable (if there is one) from the box. Set all items on a table or work area for easy access.
2. Adjust the access point antennas. Antenna configurations vary depending upon the access point model. For maximum range,

make sure the access point antennas are perpendicular to the ground, no matter where the access point will be mounted. If on a table or desk, point the antennas straight up. If mounted on a wall, point the antennas straight up, even if the access point is on its side. If mounted on the ceiling, point the antennas straight down.

3. Connect the Ethernet cable from the wired Ethernet LAN to the Ethernet connector on the back of the access point.
4. Plug the power connector into the power receptacle on the back of the access point and plug the other end into the AC outlet.
5. Plug the Ethernet cable into the access point and the other end into an Ethernet jack for the wired network.

Assign an IP Address to the Access Point

To open the access point management pages, you need the access point's IP address, which can be obtained in one of two ways:

- If your network uses a DHCP server, the server automatically assigns an IP address to the access point when it is connected to the network. Use the software applet to identify the assigned address.
- If you are not connected to a DHCP server and the access point is on the same subnet as your PC, use the access point software utility to assign an IP address to the access point.

If your access point has a serial connection, then you can also identify or set the access point's IP address by using the straight-through 9-pin serial extension cable in your access point package to connect the COM1 or COM2 port on your computer to the RS-232 serial port on the access point. Use a terminal emulator to view the access point's express setup screen and write down the IP address.

Using a DHCP Server

If your access point is configured as a DHCP client, it receives an IP address from the DHCP server when you connect it to your network.

Access Point Installation

If you have access to the DHCP server, you can use the management console to identify the IP address for the access point. If not, you need to assign the IP address using the access point software utility.

Using the Access Point Setup Utility

You must run the setup utility from a computer on the same subnet as the access point. The computer can have a wired or wireless network card and connection to the network.

Installing the Utility
The following steps are used to install the software utility on a station.

1. Put the access point software CD in the CD-ROM drive of the computer you are using to configure the access point.
2. Use Windows Explorer to view the contents of the CD. Locate and double-click the setup.exe file.
3. Follow the steps provided by the installation wizard.

Some software CDs are designed to *autorun* when inserted in the CD-ROM drive. If that is the case, the setup program wll begin automatically.

Setting the Access Point's IP Address and SSID

If your access point does not receive an IP address from a DHCP server, use the applet to assign an IP address. You can set the access point's SSID at the same time. Most applets can only change the IP address and SSID from their default settings. After these values have been changed, the utility cannot change them again unless you reset the configuration to factory defaults. To reset all system factory defaults, see the documentation of the access point.

Verifying the DHCP-Assigned IP Address

If your access point receives an IP address from a DHCP server, use the access point's applet to determine the IP address. If the client utility reports that the IP address is not one that is part of your regular network, the access point did not receive a DHCP-assigned IP address. Manually assign the IP address using the software applet that is with the access point.

Access Point Mounting

The access point is the center point of a stand-alone wireless network or the connection point to wirelessly extend the wired network. In corporate deployments, wireless users within range of an access point can roam throughout a facility while maintaining uninterrupted access to the network.

Install the access point, base station, or bridge in an area where large metal structures such as shelving units, bookcases, and filing cabinets will not obstruct radio signals being transmitted or received. Also keep it away from microwave ovens and 2 GHz cordless phones. These products may cause signal interference because they operate in the same frequency range as an access point, base station, or bridge. You can mount one to a wall or on the ceiling. The preferred location for most companies is above the drop-ceiling tile with the antenna protruding from the tiles and below the concrete ceiling.

Key Access Point Features

There are several key features of an access point to consider. These features have a direct impact on performance.

Antennas

A majority of access points designed for the enterprise or corporate environment come with omnidirectional, diversity antennas that are not removable. The access point's omnidirectional antennas provide diversity coverage for your WLAN. Diversity coverage helps maintain a clear radio signal between the access point and wireless client devices. The access point can improve signal quality by choosing the antenna that is receiving the best signal from a client device.

Some access points are equipped with *Threaded-Neill-Concelman* (TNC) connectors that enable users to connect their own antennas for special applications. The antennas provide increased gain and coverage over standard built-in antennas.

Ethernet and Serial Ports

All access ports come with an Ethernet port, and some have an Ethernet port and serial port.

Ethernet Port The access point's Ethernet port accepts a straight-through RJ-45 connector, linking the access point to the wired Ethernet LAN. Once connected to the LAN, the access point can be configured using a PC.

Serial Port If the access point has a serial port, then console access to the access point's management system is available with a serial cable. Use a straight-through serial cable to connect your computer's communications or serial port to the access point's serial port. Assign the following port settings to a terminal emulator to open the management system pages: 9600 baud, 8 data bits, no parity, 1 stop bit, and Xon/Xoff flow control.

Infrastructure Network Configuration Examples

The access point can play three distinct roles in three common wireless network configurations. The first role is also the default configuration of a root unit on a wired LAN. The other two possible roles, repeater unit and central unit in an all-wireless network, are not required for most deployments. They require specific changes to the default configuration and should be implemented after considerable research.

Root Unit on a Wired LAN

An access point connected directly to a wired LAN provides a connection point for wireless stations. If more than one access point is connected to the LAN, users can roam from one area of the office to another without losing their connection to the network. As wireless stations move out of range of one access point, they automatically connect to the network through another access point. The roaming and association process is seamless and transparent to the user (see Figure 4-1).

Repeater Unit That Extends Wireless Range An access point can be configured as a stand-alone repeater to extend the range of the wireless infrastructure or to overcome an obstacle that blocks radio communication. The repeater forwards traffic between wireless users and the wired LAN by sending packets to either another repeater or to a root access point connected to the wired LAN. The data is automatically sent through the route that provides the greatest performance for the client (see Figure 4-2).

Central Unit in an All-Wireless Network In an all-wireless network, an access point acts as a stand-alone root unit. The access point is not attached to a wired LAN; it functions as a hub linking all wireless stations together. The access point serves as the focal point for communications, increasing the security and communication range of wireless users (see Figure 4-3).

Figure 4-1
Access points as root units on a wired LAN

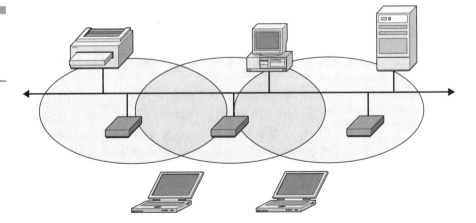

Access Point Installation

Figure 4-2
Access point as repeater

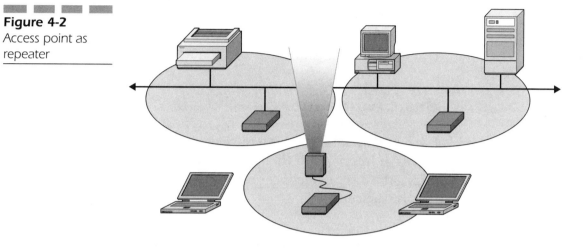

Figure 4-3
Access point as central unit in all-wireless network

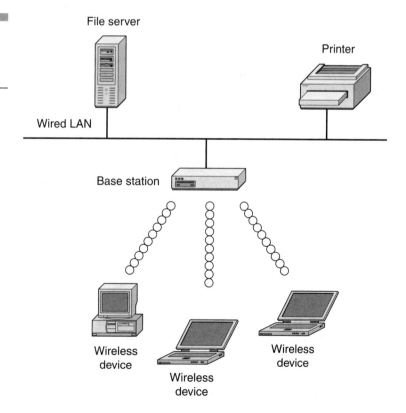

Installation Guidelines

Use the following guidelines to configure the access point.

Basic Guidelines

Because the access point is a radio device, it is susceptible to common causes of interference that can reduce throughput and range. Following the basic guidelines ensures the best possible performance of the access point.

Coverage Options The network architecture options of wireless stations and access points provide a variety of coverage alternatives and flexibility. The network can be designed to provide a wide coverage area with minimal overlap or a narrow coverage area with heavy overlap, the latter of which improves network performance and provides redundancy.

Minimal Overlap Coverage Option By arranging the access points so that the overlap in a coverage area is minimized, a large area can be covered with minimal cost. The total bandwidth available to each wireless client device depends on the amount of data each mobile station needs to transfer and the number of stations located in each cell. Seamless roaming is supported as a client device moves in and out of range of each access point, thereby maintaining a constant connection to the wired LAN. Each device in the radio network must be configured with the same SSID to provide roaming capabilities (see Figure 4-4).

Multiple Overlapping Networks Coverage Option Multiple networks can operate in the same vicinity. The 802.11 architecture enables multiple networks, each on its own radio channel, to exist in the same area with virtually no interference to each other. In this mode, each system must be configured with different SSIDs and different channels, which may prevent clients from roaming to different access points (see Figure 4-5).

Access Point Installation

Figure 4-4
Minimal overlap coverage option

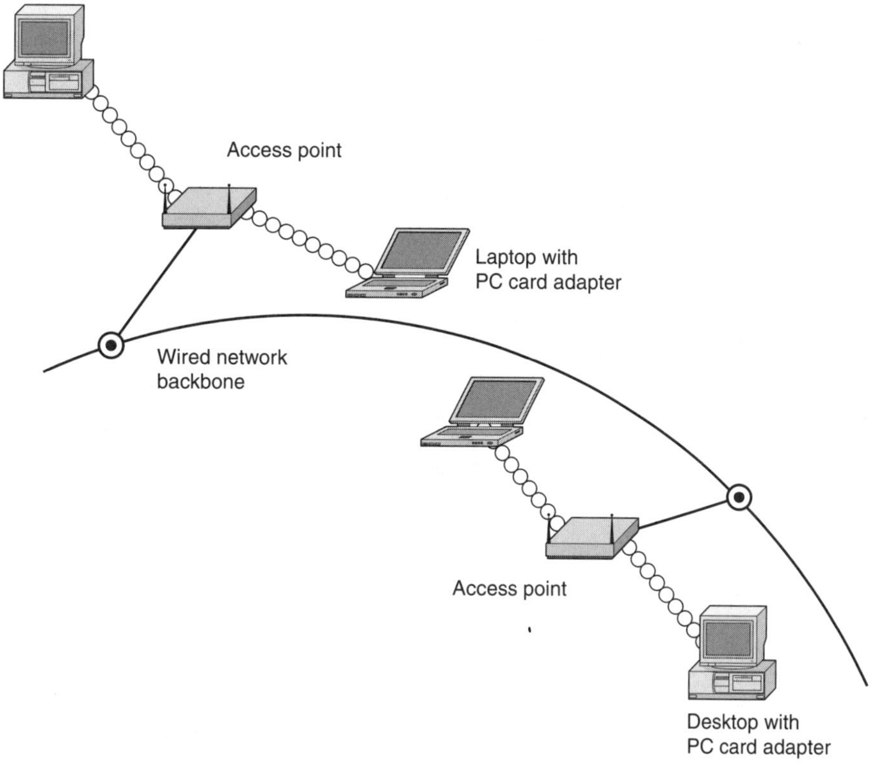

Heavy Overlap Coverage Option By arranging the access points so that the overlap in a coverage area is nearly maximized, a large number of mobile stations can be supported in the same wireless infrastructure. However, devices in overlapping coverage areas on the same frequency will detect adjacent cell traffic and delay transmissions that would cause collisions. This configuration reduces the system throughput and heavy cell overlap is not recommended for maximum system throughput. Each station must share the 11 Mbps speed with other stations on the network.

Because of the redundancy in the coverage overlap, network access is not lost if an access point fails. Upon failure of one access point, the station automatically roams to an operational access point. With this architecture, each access point in the *radio frequency* (RF) network must be configured with the same SSID to provide the roaming capability (see Figure 4-6).

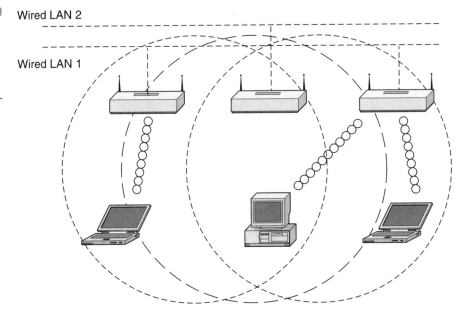

Figure 4-5
Multiple overlapping networks coverage option

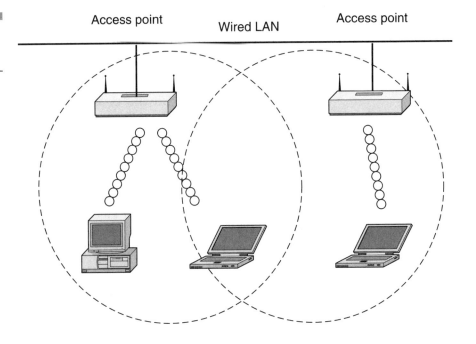

Figure 4-6
Heavy overlap coverage option

Access Point Installation

Site Surveys Because of differences in component configuration, placement, and physical environment, every network application is a unique installation. Before installing access points, you must perform a site survey to determine the optimum utilization of networking components and to maximize the range, coverage, and network performance. Site surveys can be performed with two laptop computers in an ad hoc manner for initial testing, but another site survey should be performed after the initial location for the access point has been chosen and the access point placed in the correct position before permanently mounting the access point.

Consider the following operating and environmental conditions when performing a site survey:

- ***Data rates*** Sensitivity and range are inversely proportional to data bit rates. The maximum radio range is achieved at the lowest workable data rate. A decrease in receiver threshold sensitivity occurs as the radio data increases.
- ***Antenna type and placement*** Proper antenna configuration is a critical factor in maximizing radio range. As a general rule, range increases in proportion to antenna height.
- ***Physical environment*** Clear or open areas provide better radio range than closed or filled areas. Also, the less cluttered the work environment, the greater the range.
- ***Obstructions*** A physical obstruction such as metal shelving or a steel pillar can hinder the performance of wireless devices. Avoid locating the devices in a location where there is a metal barrier between the sending and receiving antennas.
- ***Building materials*** Radio penetration is greatly influenced by the building material used in construction. For example, drywall allows a greater range than concrete blocks, while metal or steel construction is a barrier to radio signals.

Access Point Utility Software

Access points are WLAN transceivers (radio transmitter and receiver) that serve as the center point of a stand-alone wireless network or as the connection point between wireless and wired networks. Most access points use a browser-based management system, but can also be configured using a terminal emulator, a Telnet session, or the *Simple Network Management Protocol* (SNMP).

Using the Management Interfaces

A web-browser interface, a command-line interface through a terminal emulator (or a Telnet session), or an SNMP application can be used to configure the access point. Although all options are not available for all access points, the majority of access points support configuration via a web browser. The access point's management system web pages are organized the same way for the web browser and command-line interfaces.

Using the Web-Browser Interface The web-browser interface of the access point contains management pages that you use to change access point settings, upgrade and distribute firmware, and monitor and configure other wireless devices on the network.

The access point management system should be fully compatible with Microsoft Internet Explorer versions 4.0 or later and Netscape Communicator versions 4.0 or later. Earlier versions of these browsers may not be able to use all the features of the management system. The access point's web management interface can be accessed by simply typing the IP address of the access point in the address line of the browser.

Configuration

For most installations, only the basic settings require configuration. The following sections list some common settings and their functions.

Entering Basic Settings

Basic settings can be located on the Express Setup option for access point configuration. The Express Setup option is generally the first

Access Point Installation

or second option on the web management page. The settings reviewed here are derived from reviewing the web management pages of several manufacturers.

System Name The system name appears in the titles of the management system pages and in the access point's Association Table page. The system name is not an essential setting, but it helps identify the access point on your network.

The access point's MAC address appears under the system name. The MAC address is a unique serial number permanently assigned to the access point's Ethernet controller. Some access points allow the access point's MAC address to be changed. This can be useful in certain configurations, but is not usually necessary to change the MAC address.

Configuration Server Protocol Set the configuration server protocol of the access point to match the network's method of IP address assignment. In most access points, the configuration server protocol contains detailed settings for configuring the access point to work with your network's *Bootstrap Protocol* (BOOTP) or DHCP servers for the automatic assignment of IP addresses. The configuration server protocol will contain the following options:

- ***None*** Your network does not have an automatic system for IP address assignment.
- ***BOOTP*** With this protocol, IP addresses are hard-coded based on MAC addresses.
- ***DHCP*** With this protocol, IP addresses are leased for predetermined periods of time.

Default IP Address Use this setting to assign or change the access point's IP address. If DHCP or BOOTP is not enabled for your network, the IP address you enter in this field is the access point's IP address. If DHCP or BOOTP is enabled, this field provides the IP address for the access point only if no server responds with an IP address for the access point.

Default IP Subnet Mask Enter an IP subnet mask to identify the subnetwork so that the IP address can be recognized on the LAN. If

DHCP or BOOTP is not enabled, this field is the subnet mask. If DHCP or BOOTP is enabled, this field provides the subnet mask only if no server responds to the access point's DHCP or BOOTP request.

Default Gateway   Enter the IP address of your default Internet gateway here. The entry 255.255.255.255 indicates no gateway. The gateway link allows you to configure the routing page, which contains detailed settings for configuring the access point to communicate with the IP network routing system.

Radio SSID   The SSID is a unique identifier that client devices use to associate with the access point. The SSID helps client devices distinguish between multiple wireless networks in the same vicinity. Several access points on a network or subnetwork can share an SSID. The SSID can be any alphanumeric, case-sensitive entry from 2 to 32 characters long.

Role in Radio Network   This option enables you to select the role of the access point on your network. The menu contains the following options:

- ***Root Access Point***   This is a WLAN transceiver that connects an Ethernet network with wireless client stations. Use this setting if the access point is connected to the wired LAN (see Figure 4-6).
- ***Repeater Access Point***   An access point that transfers data between a client and another access point or repeater. Use this setting for access points not connected to the wired LAN (see Figure 4-7).
- ***Site Survey Client***   A wireless device that depends on an access point for its connection to the network. Use this setting when performing a site survey for a repeater access point. When you select this setting, clients are not allowed to associate.

Radio Network Optimization (Optimize Radio Network For)
You use this setting to select either preconfigured or customized settings for the access point radio:

Access Point Installation

Figure 4-7
Repeater access point

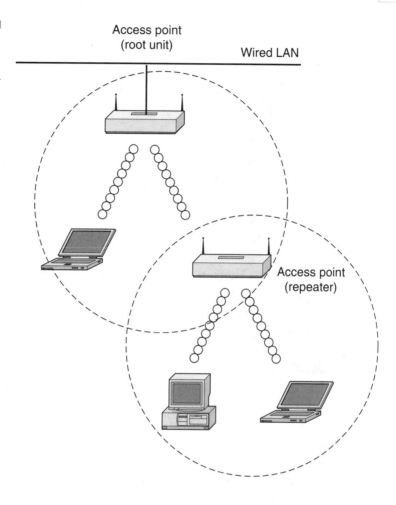

- **Throughput** Maximizes the data volume handled by the access point, but might reduce the access point's range.
- **Range** Maximizes the access point's range, but might reduce throughput.
- **Custom** The access point uses the settings you enter on the AP Radio Hardware page.

SNMP Admin. Community To use SNMP, enter a community name here. This name automatically appears in the list of users authorized to view and make changes to the access point's management system, and SNMP is enabled.

Filter Setup

This setting allows you to set up filtering to control the flow of data through the access point. You can filter data based on protocols and MAC addresses.

Protocol Filtering Protocol filters prevent or allow the use of specific protocols through the access point. You can set up individual protocol filters or sets of filters. You can filter protocols for wireless client devices, users on the wired LAN, or both. For example, an SNMP filter on the access point's radio port prevents wireless client devices from using SNMP with the access point, but does not block SNMP access from the wired LAN.

Use the Ethernet Protocol Filters page to create and enable protocol filters for the access point's Ethernet port, and use the AP Radio Protocol Filters page to create and enable protocol filters for the access point's radio port.

MAC Address Filtering MAC address filters allow or disallow the forwarding of unicast and multicast packets either sent from or addressed to specific MAC addresses. You can create a filter that passes traffic to all MAC addresses except those you specify, or you can create a filter that blocks traffic to all MAC addresses except those you specify.

MAC address filters are powerful, and you can lock yourself out of the access point if you make a mistake setting up the filters. If you accidentally lock yourself out of your access point, follow the instructions in the documentation for recovery. Most access points will have to be reset to the factory configuration, and changes made to the access point will be lost.

If you plan to disallow traffic to all MAC addresses except those you specify as allowed, put your own MAC address in the list of allowed MAC addresses first. If you plan to disallow multicast traffic, add the broadcast MAC address (ffffffffffff) to the list of allowed addresses.

Unicast packets are addressed to just one device on the network. *Multicast* packets are addressed to multiple or all devices on the subnetwork.

Access Point Installation

Client devices with blocked MAC addresses cannot send or receive data through the access point. Client devices with blocked MAC addresses disappear from the association table when the access point stops monitoring them or they associate with another access point.

Radio Configuration

This section describes how to configure the access point's radio. You use the AP Radio pages in the management system to set the radio configuration. The radio pages include the following settings:

- **AP Radio Identification**   Contains the basic locating and identity information for the access point radio port.
- **AP Radio Hardware**   Contains settings for the access point's SSID, data rates, transmit power, antennas, radio channel, and operating thresholds.
- **AP Radio Advanced**   Contains settings for the operational status of the access point's radio port. You can also use this page to make temporary changes in port status to help with troubleshooting network problems.
- **AP Radio Port**   Lists key information on the access point's radio port.

Entering Identity Information   You use the Radio Identification page to enter basic locating and identity information for the access point radio. Most pages also display the access point's MAC address, its current IP address, its current IP subnet mask, its firmware version, and its boot block version. The following settings are available for most access points:

- **Primary Port Settings**   Two options allow you to designate the access point's radio port as the primary port and select whether the radio port adopts or assumes the identity of the primary port.
- **Primary Port?**   The primary port determines the access point's MAC and IP addresses. Ordinarily, the access point's primary port is the Ethernet port, which is connected to the wired LAN, so this setting is usually set to no. Select no to set the Ethernet

port as the primary port. Select yes to set the radio port as the primary port.

- **Adopt Primary Port Identity?** Select yes to adopt the primary port settings (MAC and IP addresses) for the radio port. Select no to use different MAC and IP addresses for the radio port. Access points acting as root units adopt the primary port settings for the radio port. When you put an access point in standby mode, however, you select no for this setting. Some advanced wireless bridge configurations also require different identity settings for the radio port.

- **Default IP Address** Use this setting to assign an IP address for the radio port that is different from the access point's Ethernet IP address. During normal operation, the radio port adopts the identity of the Ethernet port. When you put an access point in standby mode, however, you assign a different IP address to the radio port. Some advanced wireless bridge configurations also require a different IP address for the radio port.

- **Default IP Subnet Mask** Enter an IP subnet mask to identify the subnetwork so that the IP address can be recognized on the LAN. If DHCP or BOOTP is not enabled, this field is the subnet mask. If DHCP or BOOTP is enabled, this field provides the subnet mask only if no server responds to the access point's request. The current IP subnet mask displayed under the setting shows the IP subnet mask currently assigned to the access point. This is the same subnet mask as the default subnet mask unless DHCP or BOOTP is enabled. If DHCP or BOOTP is enabled, this is the subnet mask used by the DHCP or BOOTP server. You can also enter this setting on the Express Setup page.

- **SSID** The SSID is a unique identifier that client devices use to associate with the access point. The SSID helps client devices distinguish between multiple wireless networks in the same vicinity. The SSID can be any alphanumeric entry from 2 to 32 characters long. You can also enter this setting on the Express Setup page.

Entering Radio Hardware Information Most access points have a Radio Hardware page and allow you to assign settings related to the access point's radio hardware. The access point Radio Hard-

Access Point Installation

ware page also contains a link to the access point Radio Data Encryption page, which you use to enter WEP settings:

- **SSID** The SSID is a unique identifier that client devices use to associate with the access point. The SSID helps client devices distinguish between multiple wireless networks in the same vicinity. The SSID can be any alphanumeric entry 2 to 32 characters long. You can also enter this setting on most Express Setup and Radio Identification pages.

- ***Allow Broadcast SSID to Associate?*** You use this setting to choose whether devices that do not specify an SSID (devices that are broadcasting in search of an access point to associate with) are allowed to associate with the access point.

  - **Yes** This is the default setting; it allows devices that do not specify an SSID to associate with the access point.

  - **No** Devices that do not specify an SSID are not allowed to associate with the access point. With no selected, the SSID used by the client device must exactly match the access point's SSID.

- ***Data Rates*** You use the data rate settings to choose the data rates that the access point uses for data transmission. The rates are expressed in Mbps. The access point always attempts to transmit at the highest rate selected. If there are obstacles or interference, the access point steps down to the highest rate that allows data transmission. For each of the four rates (1, 2, 5.5, and 11 Mbps), a drop-down menu lists three options:

  - ***Basic (default)*** Allows transmission at this rate for all packets, both unicast and multicast. At least one data rate must be set to Basic.

  - **Yes** Allows transmission at this rate for unicast packets only.

  - **No** Does not allow transmission at this rate.

The Optimize Radio Network For setting on the Express Setup page selects the data rate settings automatically. When you select Optimize Radio Network for Throughput on the Express Setup page, all four data rates are set to basic. When you select Optimize Radio Network for Range on the Express Setup page, the 1.0 data rate is set to Basic, and the other data rates are set to Yes.

The following is a list of other settings and their functions.

- ***Transmit Power***   This setting determines the power level of radio transmission. To reduce interference or to conserve power, select a lower power setting. The settings for the menu access points include 1, 5, 20, 30, 50, and 100 mW.

- ***Frag. Threshold***   This setting determines the size at which packets are fragmented (sent as several pieces instead of as one block). Enter a setting ranging from 256 to 2,338 bytes. Use a low setting in areas where communication is poor or where there is a great deal of radio interference.

- ***RTS Threshold***   This setting determines the packet size at which the access point issues a *Ready to Send* (RTS) before sending the packet. A low RTS Threshold setting can be useful in areas where many client devices are associating with the access point, or in areas where the clients are far apart and can detect only the access point and not each other. Enter a setting ranging from 0 to 2,339 bytes.

- ***Max. RTS Retries***   The maximum number of times the access point issues an RTS before stopping the attempt to send the packet through the radio. Enter a value from 1 to 128.

- ***Max. Data Retries***   The maximum number of attempts the access point makes to send a packet before giving up and dropping the packet.

- ***Data Beacon Rate***   This setting, always a multiple of the beacon period, determines how often the beacon contains a *delivery traffic indication message* (DTIM). The DTIM tells power-save client devices that a packet is waiting for them. If the beacon period is set at 100, its default setting, and the data beacon rate is set at 2, its default setting, then the access point sends a beacon containing a DTIM every 200 Kmsecs. One Kmsec equals 1,024 microseconds. This setting is generally not changed.

- ***Radio Channel***   The factory setting for various WLAN systems is radio channel 6 transmitting at 2,437 MHz. To overcome an interference problem, other channel settings are available from the drop-down menu of 11 channels ranging from

Access Point Installation

2,412 to 2,462 MHz. Each channel covers 22 MHz. The bandwidth for channels 1, 6, and 11 does not overlap, so you can set up multiple access points in the same vicinity without causing interference. Too many access points in the same vicinity creates radio congestion that can reduce throughput. A careful site survey can determine the best placement of access points for maximum radio coverage and throughput, and the best selection for the channel.

- ***Receive Antenna and Transmit Antenna***   Pull-down menus for the receive and transmit antennas offer three options:
    - ***Diversity***   This default setting tells the access point to use the antenna that receives the best signal. If your access point has two fixed (nonremovable) antennas, you should use this setting for both receive and transmit.
    - ***Right***   If your access point has removable antennas and you install a high-gain antenna on the access point's right connector, you should use this setting for both receive and transmit.
    - ***Left***   If your access point has removable antennas and you install a high-gain antenna on the access point's left connector, you should use this setting for both receive and transmit.

The access point receives and transmits using one antenna at a time, so you cannot increase the range by installing high-gain antennas on both connectors and pointing one north and one south. When the access point uses the north-pointing antenna, it would ignore client devices to the south.

Ethernet Configuration

This section describes how to configure the access point's Ethernet port. You use the Ethernet pages in the management system to set the Ethernet port configuration. The Ethernet pages include the following settings:

- ***Ethernet Identification***   Contains the basic locating and identity information for the Ethernet port.

- ***Ethernet Hardware***   Contains the setting for the access point's Ethernet port connection speed.
- ***Ethernet Advanced***   Contains settings for the operational status of the access point's Ethernet port. You can also use this page to make temporary changes in port status to help with troubleshooting network problems.
- ***Ethernet Port***   Lists key information on the access point's Ethernet port.

Entering Identity Information   You use the Ethernet Identification page to enter basic locating and identity information for the access point's Ethernet port.

Primary Port Settings   Two options allow you to designate the access point's Ethernet port as the primary port and select whether the Ethernet port adopts or assumes the identity of the primary port:

- ***Primary Port?***   The primary port determines the access point's MAC and IP addresses. Ordinarily, the access point's primary port is the Ethernet port, so this setting is usually set to yes. Select yes to set the Ethernet port as the primary port. Select no to set the radio port as the primary port.
- ***Adopt Primary Port Identity?***   Select yes to adopt the primary port settings (MAC and IP addresses) for the Ethernet port. Select no to use different MAC and IP addresses for the Ethernet port.

Some advanced bridge configurations require different settings for the Ethernet and radio ports.

Entering Ethernet Hardware Information   You use the Ethernet Hardware page to select the connector type, connection speed, and duplex setting used by the access point's Ethernet port. The Ethernet Hardware page contains one setting, Speed.

Speed   The Speed drop-down menu lists five options for the type of connector, connection speed, and duplex setting used by the port. The option you select must match the actual connector type, speed, and duplex settings used to link the port with the wired network.

Access Point Installation

The default setting, Auto, is best for most networks because the best connection speed and duplex settings are automatically negotiated between the wired LAN and the access point. If you use a setting other than Auto, make sure the hub, switch, or router to which the access point is connected supports your selection. The settings are as follows:

- *Auto* This is the default and the recommended setting. The connection speed and duplex setting are automatically negotiated between the access point and the hub, switch, or router to which the access point is connected.
- *10Base-T/Half Duplex* This is the Ethernet network connector for 10 Mbps transmission speeds over twisted-pair wire and operating in half-duplex mode.
- *10Base-T/Full Duplex* This is the Ethernet network connector for 10 Mbps transmission speeds over twisted-pair wire and operating in full-duplex mode.
- *100Base-T/Half Duplex* Ethernet network connector for 100 Mbps transmission speeds over twisted-pair wire and operating in half-duplex mode.
- *100Base-T/Full Duplex* Ethernet network connector for 100 Mbps transmission speed over twisted-pair wire and operating in full-duplex mode.

Routing Setup

You use the Routing Setup page to configure the access point to communicate with the IP network routing system. You use the page settings to specify the default gateway and to build a list of installed network route settings.

Entering Routing Settings The Routing Setup page contains the following settings:

Default Gateway Enter the IP address of your network's default gateway in this entry field. The entry 255.255.255.255 indicates no gateway.

New Network Route Settings You can define additional network routes for the access point. To add a route to the installed list, fill in the three entry fields and click Add. To remove a route from the list, highlight the route and click Remove. The three entry fields include the following:

- ***Dest Network*** Here you enter the IP address of the destination network.
- ***Gateway*** Enter the IP address of the gateway used to reach the destination network.
- ***Subnet Mask*** Enter the subnet mask associated with the destination network.

Installed Network Routes List The list of installed routes provides the destination network IP address, the gateway, and the subnet mask for each installed route.

Security Setup

One of the most important parts in the WLAN configuration is security. This section describes the types of security features you can enable on the access point.

Security Overview

The security features protect wireless communication between the access point and other wireless devices, control access to your network, and prevent unauthorized entry to the access point management system. Security features will vary by access point. These guidelines may not apply to all access points, so it is very important to choose the best available security option.

Levels of Security Security is vital for any wireless network, and you should enable the maximum-security features available on your network. The highest level of security, *Extensible Authentication Pro-*

Access Point Installation

tocol (EAP) authentication, interacts with a *Remote Authentication Dial-In User Service* (RADIUS) server on your network to provide authentication service for wireless client devices. A RADIUS server stores usernames and passwords in a database (see Figure 4-8).

If you don't enable any security features on your access point, anyone with a wireless networking device can join your network. If you enable open or shared-key authentication with WEP encryption, your network is safe from casual outsiders, but vulnerable to intruders who use a hacking algorithm to calculate the WEP key. If you enable server-based EAP authentication with *Message Integrity Check* (MIC), broadcast key rotation, and key hashing, your network is safe from all but the most sophisticated attacks against wireless security.

Encrypting Radio Signals with WEP Any wireless networking device within range of an access point can receive the access point's radio transmissions. Because WEP is the first line of defense against intruders, it is highly recommended that you use full encryption on your wireless network.

WEP encryption scrambles the communication between the access point and client devices to keep the communication private. Both the access point and client devices use the same WEP key to encrypt and unencrypt radio signals. WEP keys encrypt both unicast and multicast messages. Unicast messages are addressed to just one device on the network, while multicast messages are addressed to multiple or all devices on the network.

Figure 4-8
WLAN security levels

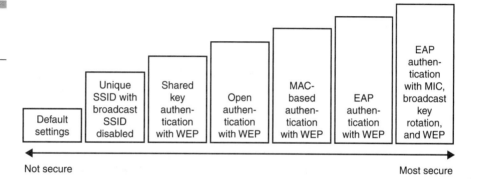

EAP authentication provides dynamic WEP keys to wireless users. Dynamic WEP keys are more secure than static WEP keys. If an intruder passively receives enough packets encrypted by the same WEP key, the intruder can perform a calculation to learn the key and use it to join your network. Because they change frequently, dynamic WEP keys prevent intruders from performing the calculation and learning the key.

Network Authentication Types Before a wireless client device can communicate on your network through the access point, it must authenticate to the access point and the network. The access point uses four authentication mechanisms or types and can use more than one at the same time:

- ***Network EAP*** This authentication type provides the highest level of security for your wireless network. By using EAP to interact with an EAP-compatible RADIUS server, the access point helps a wireless client device and the RADIUS server to perform mutual authentication and derive a dynamic unicast WEP key. The RADIUS server sends the WEP key to the access point, which uses it for all unicast data signals that it sends to or receives from the client. The access point also encrypts its broadcast WEP key (entered in the access point's WEP key slot 1) with the client's unicast key and sends it to the client. A RADIUS server or a server running RADIUS services is required on the network (see Figure 4-9).

  A wireless client device and a RADIUS server on the wired LAN use 802.1x and EAP to perform a mutual authentication through the access point. The RADIUS server sends an authentication challenge to the client. The client uses a one-way encryption of the user-supplied password to generate a response to the challenge and sends that response to the RADIUS server. Using information from its user database, the RADIUS server creates its own response and compares that to the response from the client. When the RADIUS server authenticates the client, the process repeats in reverse, and the client authenticates the RADIUS server.

  When mutual authentication is complete, the RADIUS server and the client determine a WEP key that is unique to the client

Access Point Installation

Figure 4-9
Sequence for EAP authentication

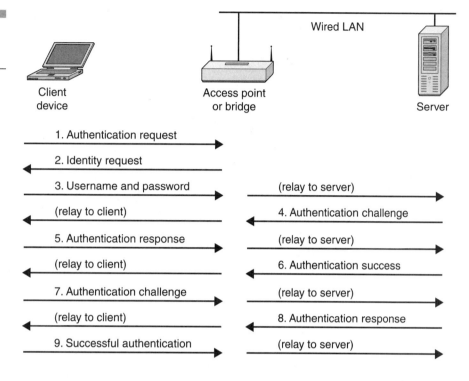

and provides the client with the appropriate level of network access, thereby approximating the level of security in a wired switched segment to an individual desktop. The client loads this key and prepares to use it for the logon session.

During the logon session, the RADIUS server encrypts and sends the WEP key, called a *session key*, over the wired LAN to the access point. The access point encrypts its broadcast key with the session key and sends the encrypted broadcast key to the client, which uses the session key to decrypt it. The client and access point activate WEP and use the session and broadcast WEP keys for all communications during the remainder of the session.

There is more than one type of EAP authentication, but the access point behaves the same way for each type; it relays authentication messages from the wireless client device to the RADIUS server and from the RADIUS server to the wireless client device.

The following security options are available for integration with EAP.

- **MAC address** The access point relays the wireless client device's MAC address to a RADIUS server on your network, and the server checks the address against a list of allowed MAC addresses. If you don't have a RADIUS server on your network, you can create the list of allowed MAC addresses on the access point. Devices with MAC addresses not on the list are not allowed to authenticate. Intruders can create counterfeit MAC addresses, so MAC-based authentication is less secure than EAP authentication. However, MAC-based authentication provides an alternate authentication method for client devices that do not have EAP capability (see Figure 4-10).

- **Open** This setting allows any device to authenticate, and then attempt to communicate with the access point. Using open authentication, any wireless device can authenticate with the access point, but the device can only communicate if its WEP keys match the access point's. Devices not using WEP do not attempt to authenticate with an access point that is using WEP. Open authentication does not rely on a RADIUS server on your network (see Figure 4-11).

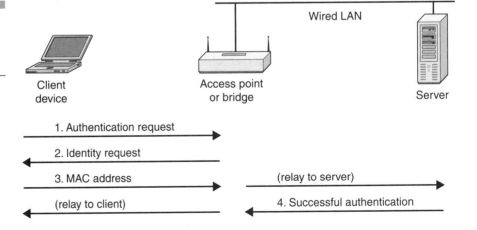

Figure 4-10
Sequence for MAC-based authentication

Access Point Installation

Figure 4-11
Sequence for open authentication

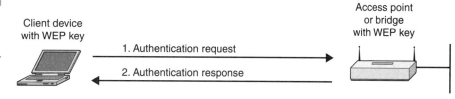

- **Shared Key** Vendors provide shared-key authentication to comply with the IEEE 802.11b standard. However, because of shared key's security flaws, it is not a recommended solution.

  During shared key authentication, the access point sends an unencrypted challenge text string to any device attempting to communicate with the access point. The device requesting authentication encrypts the challenge text and sends it back to the access point. If the challenge text is encrypted correctly, the access point allows the requesting device to authenticate. Both the unencrypted challenge and the encrypted challenge can be monitored, however, which leaves the access point open to attack from an intruder who calculates the WEP key by comparing the unencrypted and encrypted text strings. Because of this weakness, shared key authentication can be less secure than open authentication. Like open authentication, shared key authentication does not rely on a RADIUS server on your network (see Figure 4-12).

Setting Up WEP

Use the Web Management utility on the access point to manage the WEP keys. For 40-bit encryption, enter 10 hexadecimal digits; for 128-bit encryption, enter 26 hexadecimal digits. Hexadecimal digits include the numbers 0 through 9 and the letters A through F. The 40-bit WEP keys can contain any combination of 10 of these characters; the 128-bit WEP keys can contain any combination of 26 of these characters. The letters are not case sensitive.

Figure 4-12
Sequence for shared-key authentication

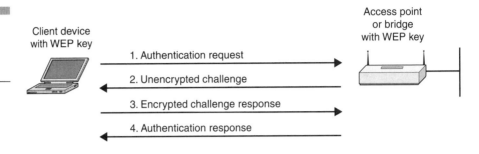

There can be up to four WEP keys. The characters you type for a key's contents appear only when you type them. After you click Apply or OK, you cannot view the key's contents.

If you enable EAP authentication, you must select key 1 as the transmit key. The access point uses the WEP key you enter in key slot 1 to encrypt multicast data signals that it sends to EAP-enabled client devices. If you enable broadcast key rotation, however, you can select any key as the transmit key, or you can enable WEP without entering any keys.

Client devices that do not use EAP to authenticate to the access point must contain the access point's transmit key in the same key slot in the client devices' WEP key lists. However, the key does not have to be selected as the transmit key in the client devices' WEP key lists.

The characters you type for the key contents appear only when you type them. After you save the changes, you cannot view the key contents. You cannot delete a WEP key, but you can write new characters over an existing key.

Enabling Additional WEP Security Features

You can enable advanced security features to protect against sophisticated attacks on your wireless network's WEP keys. This section describes how to set up and enable these features.

Enabling WEP Key Hashing WEP key hashing defends against an attack on WEP in which the intruder uses an unencrypted segment called the *initialization vector* (IV) in encrypted packets to calculate the WEP key. WEP key hashing removes the predictability that an intruder relies on to determine the WEP key by exploiting IVs. WEP key hashing protects both unicast and broadcast WEP keys.

When you enable WEP key hashing, all WEP-enabled client devices associated with the access point must support WEP key hashing. WEP-enabled devices that do not support key hashing cannot communicate with the access point.

Enabling Broadcast WEP Key Rotation EAP authentication provides dynamic unicast WEP keys for client devices, but uses static multicast keys. With broadcast, or multicast, WEP key rotation enabled, the access point provides a dynamic broadcast WEP key and changes it at the interval you select. Broadcast key rotation is an excellent alternative to WEP key hashing if your WLAN supports wireless client devices that are not Cisco devices.

When you enable broadcast key rotation, only wireless client devices using *Lightweight Extensible Application protocol* (LEAP) or EAP-TLS authentication can use the access point. Client devices using static WEP (with open, shared-key, or EAP-MD5 authentication) cannot use the access point when you enable broadcast key rotation.

You do not need to enable broadcast key rotation if you enable WEP key hashing. You can use both key rotation and key hashing, but these features provide redundant protection.

Use a short rotation interval if the traffic on your wireless network contains numerous broadcast or multicast packets.

Setting Up Open or Shared-Key Authentication Best practice recommends open authentication as preferable to shared-key authentication. The challenge queries and responses used in shared key leave the access point particularly vulnerable to intruders. This section outlines how to select open or shared-key authentication.

Setting Up MAC-Based Authentication MAC-based authentication allows only client devices with specified MAC addresses to associate and pass data through the access point. Client devices with MAC addresses not in a list of allowed MAC addresses are not allowed to associate with the access point. You can create a list of allowed MAC addresses in the access point management system and on a server used for MAC-based authentication.

Managing Firmware and Configurations

To update the firmware as it becomes available, check the web site of the access point manufacturer. Each manufacturer provides specific instructions unique to the hardware. There are just way too many combinations to list here.

Basic Troubleshooting

Troubleshooting procedures for basic problems with the access point are similar. Most common problems can be easily solved.

Checking Basic Settings

Mismatched basic settings is the most common cause of lost connectivity with wireless clients. If the access point does not communicate with client devices, check the following settings:

SSID Wireless clients attempting to associate with the access point must use the same SSID as the access point. The default SSID is tsunami.

WEP Keys The WEP key you use to transmit data must be set up exactly the same on your access point and any wireless devices with which it associates. For example, if you set WEP key 3 on your WLAN adapter to 0987654321 and select it as the transmit key, you must also set WEP key 3 on the access point to exactly the same value. The access point does not need to use key 3 as its transmit key, however.

If you use Network EAP as the authentication type, you must select key 1 as the access point's transmit key. The access point uses the WEP key you enter in key slot 1 to encrypt multicast data signals that it sends to EAP-enabled client devices. Because the access point transmits the WEP key used for multicast messages to the EAP-enabled client device during the EAP authentication process, that key does not have to appear in the EAP-enabled device's WEP key list. The access point uses a dynamic WEP key to encrypt unicast messages to EAP-enabled clients.

EAP Authentication Requires Matching 802.1x Protocol Drafts This section applies to wireless networks set up to use LEAP. If you do not use LEAP on your wireless network, you can skip this section.

Wireless client devices use EAP to log onto a network and generate a dynamic, client-specific WEP key for the current logon session. If your wireless network uses WEP without EAP, client devices use the static WEP keys entered in the Aironet Client Utilities.

If you use Network-EAP authentication on your wireless network, your client devices and access points must use the same 802.1x protocol draft. For example, if the radio firmware on the client devices that will associate with an access point or bridge is 4.16, then the access point or bridge should be configured to use Draft 8 of the 802.1x protocol.

Summary

The environment and overall network demands determine the best architecture and solution. In deploying an access point, the two most important points to consider are a site survey and the security settings. Preplanning before deployment will ensure a good, stable environment.

CHAPTER 5

Wireless Networks and Windows XP

Configuring *wireless local area networks* (WLANs) can be frustrating for the unprepared. Determining the *service set identifier* (SSID) can be difficult for new users of WLAN access points. You may have dreamed of plugging a Wi-Fi card into your laptop and having the *operating system* (OS) automatically search the airwaves and list access points for your selection of a connection. Windows XP does just that, and the Cisco Wireless *Personal Computer Memory Card International Association* (PCMCIA) Card and the Buffalo AirStation Access Point have been tested and proven to work flawlessly. The following general improvements should benefit WinXP users.

Windows XP has two editions: XP Home Edition and XP Professional Edition. Windows XP Pro enables one PC to communicate with another in a Windows NT domain, regardless of which Windows platform (Windows NT, Windows 2K, or Windows XP) either PC is using. Windows XP Pro therefore allows system administrators to maintain the system in the domain remotely. Windows XP Pro supports dual processors; many applications currently available do not. Each OS version has its own upgrade path. For example, current users of Windows NT 4 or Windows 2K will upgrade to XP Pro. Users upgrading from Windows 98, Windows 98 SE, or Windows ME can choose Windows XP Home or Windows XP Pro; however, there is no upgrade from Windows 95 directly to Window XP.

Windows XP includes the best support for a WLAN setup of any current OS. Insert a WLAN card into your PC, and you will find a device icon in the lower-right corner of the screen, indicating the driver has automatically loaded. The setup wizard pops up, and by clicking Next, you automatically install the card. You will be asked to select an access point for connection within the "Connect to wireless network" window. Select from the list of possible access points, simply type in a network key (*Wired Equivalent Privacy* [WEP]) if necessary, and then press Connect. Once set, the system is ready to go. Windows XP will seamlessly switch from LAN to LAN without any necessary intervention on your part.

The first time you encounter a new network, you are prompted for its WEP key. Thereafter, Windows stores settings and connects automatically. If you have ever tired of switching between WLANs at home and work, you are more than familiar with the routine of multiple reboots, manually changing *Extended Service Set IDs* (ESS-IDs) and WEP keys, and locating lost *Internet Protocol* (IP) address leases. Windows XP makes WLAN configuration easier and automates several steps required by the user.

To maintain security, Windows XP integrates the stealth firewall in the OS. When an offending PC scans the IP of your system, the firewall sends no response. Any hacker attempting to tap the network will simply not find a PC there. To minimize compatibility issues with Windows file and printer sharing, the Windows XP firewall can be disabled by default.

Microsoft Windows treats a wireless networking card just as any other device. Whether it be on a desktop or laptop, the process for installing and configuring a wireless networking card is the same straightforward process. Simply physically insert or install the new card in the computer; then install the appropriate drivers for the correct combination of hardware and software. If the user is on a fixed wireless system, such as a desktop that remains in one microcell, the solution is simply to configure the SSID and WEP key.

To provide security beyond the 802.11b WEP standard, Microsoft worked closely with the *Institute of Electrical and Electronics Engineers* (IEEE) standards committee and other networking manufactures to define the emerging 802.1X standard. This is a draft for port-based network access used to provide authentication for network access. It uses the physical characteristics of the LAN to authenticate devices attached to a LAN. Originally designed for wired Ethernet networks, the 802.1X security standard has found its way to WLANs.

For the case of the WLAN, the access point will provide authentication to the network using *Remote Authentication Dial-In User Service* (RADIUS) to authenticate the client. Initial communication is initiated through a logical or open port or channel on the access point for sending and validating credentials and for receiving keys to access the network through a logical or controlled port. Once the keys are available to the access point and client, they enable data to be encrypted and authenticated by the access point. This solution adds key management to the 802.11 security scheme.

The following are the general authentication steps for 802.11X:

1. When a new client enters the range of the access point, the access point issues the client an authentication challenge. Without a valid key, the access point restricts data flow.
2. When the client receives the challenge, it replies with its identification. The access point forwards the identification information to a RADIUS server for authentication and waits for a response.
3. The RADIUS server requests the credentials for the client, and the client sends its information to the RADIUS server through the access point's open port.
4. The RADIUS server assumes validity, validates the client's identification, and sends an authentication key to the access point. The authentication key is encrypted so only the access point can interpret it.
5. The access point uses the authentication key to securely transmit the appropriate keys to the client that includes a unicast session key and a global session key for multicasts by the client.
6. The client is sometimes asked to reauthenticate periodically to maintain security.

Easing the Security Burden with RADIUS

The 802.1X approach builds on the widespread use of RADIUS for authentication. Authentication credentials in a RADIUS server are generally stored in a secured database, and a RADIUS server can query the local database or forward the authentication request to another server.

When a RADIUS server authenticates the machine and sends a message back to the access point, the access point allows a secured connection to be created between the client and access point by communicating with session and global keys.

In the real world, the process looks similar to this:

1. A salesperson is traveling and has a laptop with an 802.11 network card. The salesperson is waiting at the airport and has some extra time and turns on the laptop.

2. The laptop powers up, and the wireless network card finds an available wireless network access point and associates with it. If the acess point is using a SSID, the user must enter SSID, unless the laptop is Windows XP. In that case, Windows XP will automatically determine the SSID.

3. The laptop sends the credentials (username and password) to the access point for authentication.

4. The access point forwards the request to a RADIUS server, and the server determines that the username is from a company that is allowed to connect to the wireless network. That is, the salesperson's company is paying to enables their users to use the wireless network.

5. The RADIUS server forwards the request to the company's server to verify the credentials. Once the company's server responds with a yes, the access point enables the laptop to access the wireless network (see Figure 5-1).

To allow for this level of security and functionality, Microsoft Windows XP is providing a client-side implementation and enhancing Windows RADIUS server and *Internet Authentication Server* (IAS) to support wireless network authentication. In addition to enhancing the operating system, Microsoft has worked closely with the 802.11 device vendors to support this functionality in the wireless *network interface card* (NICs) and in access points.

Windows 2000 includes enhancements that enable it to detect the availability of a network, wired or wireless, and allow connecting with minimal user configuration. Windows XP provides additional functionality not found in Windows 2000. Windows XP supports the truly mobile nature of wireless networking.

Windows 2000 uses media sense capabilities (detecting an attached network) to control the configuration of the network stack and inform the user when the network is unavailable. In Windows XP, the network stack is used to enhance the wireless roaming

Figure 5-1
An example public access scenario

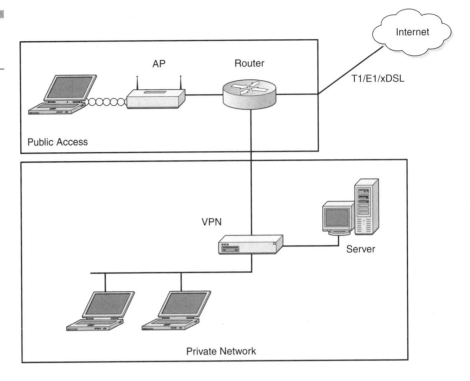

experience by detecting a move to a new access point, forcing reauthentication to ensure appropriate network access, and detecting changes in an IP subnet so an appropriate address can be used to gain resource access.

Multiple IP address configurations (*Dynamic Host Configuration Protocol* [DHCP] assigned or static address) can be available on a Windows XP system, and Windows XP can choose the appropriate configuration automatically. If the IP address of the Windows XP changes, Windows automatically reconfigures itself, if appropriate. By using the Windows Sockets extensions, network-aware applications (Internet browsers, firewalls, and so on) become aware of connectivity and adjust automatically. The auto-sensing and reconfiguration negates the need for mobile IP in a wireless network and solves most user issues when roaming between networks. As always, applications are completely unaware of the changes taking place and continute to work in the normal state.

When stations roam from access points, certain crucial information must be updated on the access point and client station, including IP routing information for packet delivery and messaging. Rather than go through the process of disassociation and reassociation, an access point can forward the information to the new access point. To achieve this, no protocol has been defined, but several WLAN vendors have joined together to develop and propose the *Inter-Access Point Protocol* (IAAP).

Zero Configuration Required

Windows XP has a "Wireless Configuration Zero" service that configures the wireless network card with a wireless network. The wireless network card will scan for available networks and forward that information to Windows XP. The wireless network card or the drivers do not require additional user configuration or additional overhead. The network card and driver simply pass on a few additional *object identifiers* (OIDs) used to query the network adapter and set device and driver behavior. If multiple WLANs are covering the same area, users can configure a preferred network order, and the machine will seek active LAN connections in that order. With Windows XP, it is possible to limit association to only the configured, preferred wireless networks.

If no access point is located, Windows XP will automatically configure the wireless network card into ad hoc networking by default. To prevent this from occurring at undesired times, the wireless network card can be set to be disabled or forced into ad hoc mode if no infrastructure network is found.

The zero configuration enhancements in Windows XP are designed to take the configuration chore away from the user. In addition to the zero configuration enhancements, additional security enhancements attempt to locate other wireless networks if authentication fails with the current wireless network.

Windows XP and Wireless Security

Microsoft has teamed up with Cisco Systems, Dell Computer, Intel, and other leading computer manufacturers to improve security for wireless network connections at the office, home, and public places. The new proposed security standard, 802.1X, will fix many of the security vulnerabilities recently found on wireless networks based on the wireless 802.11b standard. The security flaws discovered in 802.11b could let hackers pick up and alter wireless data communications. The new wireless security standard will ease concerns about security and help increase deployments of wireless networks in the enterprise and for the home. The new proposed standard is due to be approved by the IEEE in the near future.

With the new standard and support from Microsoft, Cisco, and other wireless card manufactures, users have a way to connect to a wireless corporate network or a wireless public network with no configuration changes required. For the most secured wireless network, administrators may consider running a *virtual private network* (VPN) over the wireless network.

Installation and Configuration Guide

When installing the componets of a wireless network, the access point, and the wireless NICs, it is helpful to understand the components and their function. Knowing what a LED flashing in a particular pattern can often help in understanding and troubleshooting the problem.

Wireless Adapters

The WLAN network cards (wireless NICs) are peripherals that provide wireless data communications between fixed, portable, or mobile devices and other wireless devices or a wired network infrastructure. Most client adapters are fully compatible when used in devices supporting *Plug-and-Play* (PnP) technology.

The primary function of the client adapters is to transfer data packets transparently through the wireless infrastructure. The adapters operate similarly to a standard network product, except that the cable is replaced with a radio connection. No special wireless networking functions are required, and all existing applications that operate over a network can operate using the adapters.

There are several types of wireless network cards for various devices. The most prevalent devices in the market are as follows:

- ***PC card client adapter (also referred to as a PCMCIA card)*** A PC card radio module that can be inserted into any device equipped with an external Type II or Type III PC card slot. Host devices can include laptops (notebook computers), *personal digital assistants* (PDAs), and handheld or portable devices.

- ***Perhipheral Component Interconnect (PCI) card client adapter*** A client adapter card radio module that can be inserted into any device equipped with an empty PCI expansion slot, such as a desktop computer.

- ***ISA card client adapter*** A client adapter card radio module that can be inserted into any device equipped with an empty ISA expansion slot, such as a desktop computer. Most older computers use ISA expansion slots.

- ***Mini-PCI card client adapter (also referred to as a mini-PCI card)*** A client adapter card radio module that can be inserted into any device equipped with an internal Type IIIA mini-PCI card slot, such as a laptop computer. Most mini-PCI cards are built into the laptop or notebook computer.

Hardware Components of the Client Adapter The wireless NIC has three major hardware components: a radio, a radio antenna, and various *light-emitting diodes* (LEDs) indicating power and status. The number, type, and information displayed by the LEDs vary by manufacturer.

Radio The client adapter contains a radio that operates in the 2.4 GHz band. The radio transmits data without the need of wires, is mounted on the surface of the card, and is generally a part of the network card itself.

Radio Antenna Virtually all PC cards have an integrated, permanently attached directional diversity antenna. The benefit of the diversity antenna system is improved coverage. The system works by enabling the card to switch and sample between its two antenna ports in order to select the optimum port for receiving data packets. As a result, the card has a better chance of maintaining the *radio frequency* (RF) connection in areas of interference. The antenna is housed within the section of the card that hangs out of the PC card slot when the card is installed.

Most PCI cards are shipped with a dipole antenna that attaches to the card's antenna connector. However, other types of antennas may be used. PCI cards can be operated through only one port unless the card supports diversity antennas. Most cards support the diversity antenna function and can be easily differentiated from the non-diversity enabled cards. Cards that support diversity antenna have two connectors for the antenna protruding from the back of the PCI card.

LEDs PC card and PCI client adapters come with one or more LEDs that indicate the status of the adapter or to convey messages or card status. The number of LEDs and color varies from manufacturers and in some cases, models. The number of LEDs can vary from 1 to 3, and some common colors are yellow, green, and orange.

Software Components of Client Adapters The client adapter has three major software components: radio firmware, a driver, and a client utility.

Radio Firmware The firmware, which is contained in the client adapter's flash memory, controls the adapter's radio. The client adapter is shipped with the firmware installed; however, a more recent version of the firmware may be available from the manufacturer's web site as manufacturers release updates regularly.

Driver The driver provides an interface between a computer running a Windows operating system and the client adapter, thereby enabling Windows and the applications it runs to communicate with the adapter. The driver is provided on the CD that shipped with the client adapter and must be installed on the client computer before the adapter can be used. The CD has the latest version of the driver available at the time of shipment; however, a more recent version of the driver may be available from the manufacturer's web site.

Client Utility The client utility is an optional applet that interacts with the radio firmware to adjust client adapter settings and display information about the adapter. The utility applet is provided on the CD that shipped with the client adapter and should be installed before the adapter is used. The software CD has the latest version of the utility available at the time of shipping; however, a more recent version of the applet may be available from the manufacturer's web site. Utilities are manufacturer specific and cannot be interchanged.

If your computer is running Windows XP, however, you can configure your client adapter through Windows instead of having to use the utility. Windows XP supports a large number of client network cards.

Client Adapter and Possible Network Configurations

The client adapter can be used in a variety of network configurations. In some configurations, access points provide connections to your network or act as repeaters to increase the wireless communication range, which is based on how you configure your wireless network.

Ad Hoc WLAN An ad hoc (or peer-to-peer) WLAN is the simplest WLAN configuration. In a WLAN using an ad hoc network configuration, all devices equipped with a client adapter can be linked together and communicate directly with each other.

Infrastructure LAN with Wired LAN Connectivity A micro wireless network can be created by placing two or more access points on a LAN with workstations accessing a wired LAN through the access points. This configuration is useful with portable or mobile stations because it allows them to be directly connected to the wired network even while moving from one micro wireless domain to another. This process is transparent, and the connection to the file server or host is maintained without disruption. The mobile station stays connected to an access point as long as it can. However, once the transfer of data packets needs to be retried or beacons are missed, the station automatically searches for and associates with another access point. This process is referred to as *seamless roaming*, which will be discussed later in the book. Roaming is a function of the hardware, firmware, and OS present on the computer.

Preparing for Installation

Installing the Client Adapter is a straighforward process. It consists of installing the hardware and drivers. Following the steps listed provides a generic guideline for the installation process.

Unpacking the Client Adapter

1. Open the shipping container and carefully remove the contents.
2. Return all packing materials to the shipping container and save it.
3. The adapter and CD are the items necessary for installation and configuration, so be sure to keep them in a safe place.

System Requirements

In addition to the items shipped with the client adapter, you will also need the following in order to install and use the adapter:

- One of the following computing devices running Windows 2000, XP, or CE:
    - A laptop, notebook, or portable/handheld device equipped with a Type II or Type III PC card slot if the PC card will be installed in a laptop or notebook computer

- Desktop PC equipped with an empty PCI expansion slot if the adapter needs to be installed in a desktop
- Laptop or other computing device with an embedded mini-PCI card
■ A Phillips screwdriver for PCI card installation
■ The following information from your system administrator:
  - The workstation's client name
  - The protocols necessary to bind to the client adapter
  - The case-sensitive SSID for your RF network
  - If your computer is not connected to a DHCP server, the IP address, subnet mask, and default gateway address of the computer
  - The WEP keys of the access points with which your client adapter will communicate if your wireless network uses static WEP for security
  - The username and password for your network account
  - The username and password for your RADIUS server account if your wireless network uses *Lightweight Extensible Authentication Protocol* (LEAP) or EAP-MD5 authentication

Site Requirements

Site requirements for an infrastructure network are similar to those of a wired network. Because of differences in component configuration, placement, and physical environment, every network application is a unique installation.

For Infrastructure Devices Before you install any wireless infrastructure devices (such as access points, bridges, and base stations, which connect your client adapters to a wired LAN), a site survey must be performed to determine the optimum placement of these devices to maximize range, coverage, and network performance. A site survey should be used to determine the best placement for infrastructure devices within a wireless network. Wireless infrastructure devices, such as an access point, must be installed before the client adapters in devices.

For Client Devices Because the client adapter is a radio device, it is susceptible to RF obstructions and common sources of interference that can reduce throughput and range. Follow these guidelines to ensure the best possible performance:

- Install the client adapter in an area where large metal structures, such as shelving units, bookcases, and filing cabinets, will not obstruct radio signals to and from the client adapter.
- Install the client adapter away from microwave ovens. Microwave ovens operate in the same frequency range as the client adapter and can cause signal interference.

Be sure to consider obstacles unique to your environment, industry, or business. For example, if you deploy a wireless network in a hospital environment, attention must be paid to make sure that the wireless network does not cause any RF interference with other life-critical equipment. Ensuring devices are *united laboratories* (UL) certified is a good indicator.

Installation of Client Adapters

The next step is to install the client adapter in the computing device. The following guidelines provide a good framework to follow.

Installing the Driver

This procedure is meant to be used the first time the driver is installed on a computer running Windows 2000 or XP. If a client adapter driver is already installed on your computer, follow the manufacturer's instructions to upgrade to a new driver.

The driver you use for your client adapter depends on which operating system your computer is running. Before you begin the driver installation process, make sure you have the installation disks for your computer's operating system nearby. Some operating system files may be needed to complete the driver installation.

Wireless Networks and Windows XP

Installing the Driver for Windows 2000 To install a driver in a Windows 2000 system, follow these steps:

1. Insert the client adapter into your computer. Windows 2000 automatically detects the client adapter and briefly opens the Found New Hardware Wizard window, which indicates that the wizard will help you to install the driver.

2. Click Next. Another window opens and asks what you want the wizard to do.

3. Select "Display a list of the known drivers for this device so that I can choose a specific driver" and click Next.

4. Click Have Disk.

5. Insert the adapter's CD or a floppy disk containing the latest driver into your computer, unless you are installing the driver from your computer's hard drive.

6. Enter or browse to the path where the driver is located (CD, floppy disk, or hard drive) and click OK. The wizard finds the installation files and displays the search results.

7. When the client adapter driver is displayed, click Next to copy the required files.

8. When you receive a message indicating that Windows has finished the installation, click Finish.

9. Remove the CD or floppy disk (if installed).

10. Double-click My Computer, Control Panel, and System.

11. In the System Properties window, click the Hardware tab.

12. Click Device Manager.

13. In the Device Manager window, double-click Network Adapters.

14. Right-click the wireless LAN adapter.

15. Click Properties.

16. In the client adapter Properties window, click the Advanced tab.

17. In the Advanced window, select Client Name. Type your computer's unique client name, which can be obtained from your system administrator, in the Value dialog box.

18. Select SSID. Type your RF network's (case-sensitive) SSID, which can be obtained from your system administrator, in the Value dialog box.
19. Click OK.
20. If your computer is not connected to a DHCP server and you plan to use TCP/IP, follow these steps:
    a. Double-click My Computer, Control Panel, and Network and Dial-up Connections.
    b. Right-click Local Area Connection.
    c. Click Properties, Internet Protocol (TCP/IP), and Properties.
    d. Click "Use the following IP address" and enter the IP address, subnet mask, and default gateway address of your computer (which can be obtained from your system administrator). Click OK.
    e. In the Local Area Connection Properties window, click OK.
21. If you are prompted to restart your computer, click Yes. The driver installation is complete.

Installing the Driver for Windows XP Windows XP comes with most drivers that are installed automatically the first time you insert a client adapter. To upgrade to the driver provided on the adapter's CD, follow these steps:

1. Insert the client adapter into your computer. The instructions vary by operating system and are different for PC and PCI cards.
2. Double-click My Computer, Control Panel, and System.
3. Click the Hardware tab and Device Manager.
4. Double-click Network Adapters and Wireless LAN Adapter.
5. Click the Driver tab.
6. Click Update Driver. The Welcome to the Hardware Update Wizard screen appears.
7. Select the "Install from a list or specific location (Advanced)" option and click Next.

Wireless Networks and Windows XP

8. When prompted to choose your search and installation options, select "Don't search. I will choose the driver to install" and click Next.

9. When prompted to select a network adapter to install, click Have Disk. The Install From Disk screen appears.

10. Insert the Wireless Adapter's CD or a floppy disk containing the latest driver into your computer, unless you are installing the driver from your computer's hard drive.

11. Click Browse, find the location of the driver (on your CD, floppy disk, or computer's hard drive), and click Open. The installation wizard finds the driver file. Click OK on the Install From Disk screen.

12. The Select Network Adapter screen reappears. Select Wireless LAN Adapter and click Next.

13. The installation wizard copies the driver files from the CD, floppy disk, or computer's hard drive. When the installation is complete, click Finish.

14. Click Close on the wireless adapter's Properties screen and exit the Control Panel.

15. Double-click Control Panel and Network Connections.

16. Right-click Wireless Network Connection.

17. Click Properties, Configure, and the Advanced tab.

18. In the Advanced window, select Client Name. Type your computer's unique client name, which can be obtained from your system administrator, in the Value dialog box.

19. Select SSID. Type your RF network's (case-sensitive) SSID, which can be obtained from your system administrator, in the Value dialog box.

20. Click OK.

21. If your computer is not connected to a DHCP server and you plan to use TCP/IP, right-click Wireless Network Connection and click Properties. Select Internet Protocol (TCP/IP) and click Properties. Select Use the following IP address and enter the IP

address, subnet mask, and default gateway address of your computer. Click OK.

22. The driver installation is complete. Now you must decide whether to configure your client adapter through Windows XP or your network card's *Client Utilities*.

23. Perform the following:

    - If you are planning to configure your client adapter through ACU instead of through Windows XP, follow these steps:
        a. Double-click My Computer, Control Panel, and Network Connections.
        b. Right-click Wireless Network Connection and click Properties.
        c. Select the Wireless Networks tab.
        d. Deselect the "Use Windows to configure my wireless network settings" checkbox.
        e. Follow the instructions provided by the manufacturer to install the wireless adapter applet.

Installing the Adapter Software Utility

After you have installed the appropriate driver for your computer's operating system, follow these steps to install the adapter utility:

1. Close any Windows programs that are running.
2. Perform one of the following:
    - If you are installing an adapter utility from the wireless adapter's CD that shipped with the client adapter, follow these steps:
        a. Insert the CD into your computer's CD-ROM drive.
        b. Select Start, Run, and enter the following path (where D is the letter of your CD-ROM drive): locate the setup.exe file and double-click the file. The Adapter Utility Setup screen and the Installation Wizard appear. Follow the wizard instructions.

Wireless Networks and Windows XP

- If your computer needs to be rebooted, select "Yes, I want to restart my computer now" or "No, I will restart my computer later," remove the CD or floppy disk (if installed), and click Finish.

Verifying Installation

To verify that you have properly installed the driver and adapter utility and minimally configured your client adapter, check the client adapter's LEDs. If the installation was successful, the client adapter's LED blinks with activity.

Configuring the Client Adapter

Once the physical installation is complete, configure the adapter's parameters to match the settings used by the wireless network or other wireless computing devices. In most cases, the default settings (except for the security settings) are acceptable.

Setting RF Network Parameters

The default RF parameters are acceptable for a majority of implementations. But if you need to change any of the settings, an understading of the function of the settings will be helpful.

- ***Data Rate*** Specifies the rate at which your client adapter should transmit or receive packets to or from access points (in infrastructure mode) or other clients (in ad hoc mode). Auto Rate Selection is recommended for infrastructure mode; setting a specific data rate is recommended for ad hoc mode. Options vary by manufacturer but include Auto Rate, 1 Mbps Only, 2 Mbps Only, 5.5 Mbps Only, or 11 Mbps Only. Your client adapter's data rate must be set to Auto Rate or must match the data rate of the access point (in infrastructure mode) or the other clients (in ad hoc mode) with which it is to communicate. Otherwise, your client adapter may not be able to associate to them. Auto Rate Selection is the recommended setting.

- *Auto Rate Selection*   Uses the 11 Mbps data rate when possible, but drops to lower rates when necessary.
- *1 Mbps Only*   Offers the greatest range but the lowest throughput.
- *2 Mbps Only*   Offers less range but greater throughput than the 1 Mbps Only option.
- *5 Mbps Only*   Offers less range but greater throughput than the 2 Mbps Only option.
- *11 Mbps Only*   Offers the greatest throughput but the lowest range.

- **Use Short Radio Headers**   Selecting this checkbox sets your client adapter to use short radio headers. However, the adapter can use short radio headers only if the access point is also configured to support them and is using them. If any clients associated with an access point are using long headers, then *all* clients in that cell must also use long headers, even if both this client and the access point have short radio headers enabled. Short radio headers improve throughput performance; long radio headers ensure compatibility with clients and access points that do not support short radio headers. This parameter is referred to as *preambles* on the access point screens.

- **World Mode**   Selecting this checkbox enables the client adapter to adopt the maximum transmit power level and the frequency range of the access point to which it is associated, provided the access point is also configured for World Mode. This parameter is available only in infrastructure mode and is designed for users who travel between countries and want their client adapters to associate to access points in different regulatory domains. When World Mode is enabled, the client adapter is limited to the maximum transmit power level allowed by the country of operation's regulatory agency.

- **Channel**   Specifies which frequency your client adapter will use as the channel for communications. These channels conform to the IEEE 802.11 standard for your regulatory domain. In infrastructure mode, this parameter is set automatically by some network adapters and cannot be changed. Some client adapters listen to the entire spectrum, select the best access point to associate to, and use the same frequency as that access point. In ad hoc mode, the channel of the client adapter must be set to match the channel used by the other clients in the wireless network.

- **Transmit Power** Defines the power level at which your client adapter transmits. This value must not be higher than that allowed by your country's regulatory agency. Reducing the transmit power level conserves battery power but decreases radio range.

- **Data Retries** Defines the number of times a packet will be re-sent if the initial transmission is unsuccessful. The range is from 1 to 128. If your network protocol performs its own retries, set this to a smaller value than the default. This way notification of a "bad" packet is sent up the protocol stack quickly so the application can retransmit the packet, if necessary.

- **Fragment Threshold** Defines the threshold above which an RF data packet will be split up or fragmented. If one of those fragmented packets experiences interference during transmission, only that specific packet would need to be re-sent. Throughput is generally lower for fragmented packets because the fixed packet overhead consumes a higher portion of the RF bandwidth. The range is from 256 to 2,312.

Setting Advanced Infrastructure Parameters

The advanced infrastructure parameters can be set only if the client adapter has been set to operate on an infrastructure network. The parameters are as follows:

- **Antenna Mode (Receive)** Specifies the antenna that your client adapter uses to receive data.

- **PC card** The PC card's integrated, permanently attached antenna operates best when used in Diversity Mode. Diversity

Mode enables the card to use the better signal from its two antenna ports. The options here are Diversity (Both), Right Antenna Only, and Left Antenna Only.

- ***LM card***   The LM card is shipped without an antenna; however, an antenna can be connected through the card's external connector. If a snap-on antenna is used, Diversity Mode is recommended. Otherwise, select the mode that corresponds to the antenna port to which the antenna is connected. The options are Diversity (Both), Right Antenna Only, and Left Antenna Only.
- ***PCI client adapter***   The PCI client adapter must use the Right Antenna Only option.
- ***Mini-PCI card***   The mini-PCI card, which can be used with one or two antennas, operates best in Diversity Mode. Diversity Mode enables the card to use the better signal from its two antenna connectors. The options are Diversity (Both), Right Antenna Only, and Left Antenna Only.

- ***Antenna Mode (Transmit)***   Specifies the antenna that your client adapter uses to transmit data. See the Antenna Mode (Receive) parameter earlier for information on the options available for your client adapter.

- ***Specified Access Point***   Specifies the MAC addresses of up to four preferred access points with which the client adapter can associate. If the specified access points are not found or the client adapter roams out of range, the adapter may associate to another access point. You can enter the MAC addresses of the access points in the edit boxes or choose not to specify access points by leaving the boxes blank. This parameter should be used only for access points that are in repeater mode. For normal operation, leave these fields blank because specifying an access point slows down the roaming process.

- ***RTS Threshold***   Specifies the size of the data packet that the low-level RF protocol issues to a *request-to-send* (RTS) packet. Setting this parameter to a small value causes RTS packets to be sent more often. When this occurs, more of the available bandwidth is consumed and the throughput of other network packets is reduced, but the system is able to recover faster from interference or collisions, which may be caused from a high multipath environment characterized by obstructions or metallic surfaces.

- **RTS Retry Limit** Specifies the number of times the client adapter will resend a RTS packet if it does not receive a *clear-to-send* (CTS) packet from the previously sent RTS packet. Setting this parameter to a large value decreases the available bandwidth whenever interference is encountered but makes the system more immune to interference and collisions, which may be caused from a high multipath environment characterized by obstructions or metallic surfaces. The range is from 1 to 128.

Setting Advanced Ad Hoc Parameters

You can set advanced ad hoc parameters only if your client adapter has been set to operate in an ad hoc network.

Setting Network Security Parameters

The Network Security section enables you to set parameters that control how the client adapter associates with an access point, authenticates to the wireless network, and encrypts and decrypts data. To access this screen, select the Network Security tab from the Properties screens.

This screen is different from the other Properties screens in that it presents several security features, each of which involves a number of steps. In addition, the security features themselves are complex and need to be understood before they are implemented. In order to use any of the security features, both your client and the access point to which it will associate must be set appropriately. Table 5-1 shows a summary of the security settings and a matrix of the best settings for the OS and security settings. (OR) A summary of the security settings and the OS for the best use is listed for comparison.

Table 5-1

Client and Access Point Security Settings

Security Feature	Client Setting	Access Point Setting
Static WEP with open authentication	Use static WEP keys, open authentication enabled, and a WEP key created	Open authentication and WEP enabled and a WEP key created
Static WEP with shared key authentication	Use static WEP keys, shared key authentication enabled, and a WEP key created	Shared key authentication and WEP enabled and a WEP key created
LEAP authentication	LEAP enabled	Network-EAP enabled
EAP-TLS authentication If using ACU to configure a card	Host-based EAP enabled in ACU and a Smart Card or other certificate enabled in Windows XP	Network-EAP enabled
If using Windows XP to configure a card	Smart Card or other certificate enabled	Require EAP and open authentication enabled
EAP-MD5 authentication If using ACU to configure a card	Host-based EAP enabled in ACU and MD5-Challenge enabled in Windows XP	Network-EAP enabled
If using Windows XP to configure a card	MD5-Challenge enabled	Require EAP and open authentication enabled

Overview of Security Features

When you use your client adapter with Windows 2000, you can protect your data as it is transmitted through your wireless network by encrypting it through the use of WEP encryption keys. With WEP encryption, the transmitting device encrypts each packet with a WEP key, and the receiving device uses that same key to decrypt each packet.

The WEP keys used to encrypt and decrypt transmitted data can be statically associated with your adapter or dynamically created as part of the EAP authentication process. Dynamic WEP keys with EAP offer a higher degree of security than static WEP keys.

WEP keys, whether static or dynamic, are either 40 or 128 bits in length; 128-bit WEP keys offer a greater level of security than 40-bit WEP keys.

Static WEP Keys Each device within the wireless network can be assigned up to four static WEP keys. If a device receives a packet that is not encrypted with the appropriate key (as the WEP keys of all devices that are to communicate with each other must match), the device discards the packet and never delivers it to the intended receiver.

Static WEP keys are write-only and temporary; therefore, they cannot be read back from the client adapter, and they are lost when power to the adapter is removed or the Windows device is rebooted. Although the keys are temporary, you do not need to re-enter them each time the client adapter is inserted or the Windows device is rebooted. This is because the keys are stored (in an encrypted format for security reasons) in the registry of the Windows device. When the driver loads and reads the client adapter's registry parameters, it also finds the static WEP keys, unencrypts them, and stores them in volatile memory on the adapter.

The Network Security screen enables you to view the current WEP key settings for the client adapter and then to assign new WEP keys or overwrite existing WEP keys, as well as to enable or disable static WEP.

EAP (with Static or Dynamic WEP Keys) The new standard for WLAN security, as defined by the IEEE, is called 802.1X for 802.11, or simply 802.1X. An access point that supports 802.1X and its protocol, EAP, acts as the interface between a wireless client and an authentication server, such as a RADIUS server, to which the access point communicates over the wired network.

Two 802.1X authentication types can be selected in ACU for use with Windows operating systems:

- ***Host-based EAP*** Selecting this option enables you to use any 802.1X authentication type for which your operating system has built-in support. For example, Windows XP has built-in support for both EAP-TLS and EAP-MD5.
  - ***EAP-TLS*** EAP-TLS is enabled or disabled through the operating system and uses a dynamic, session-based WEP key, which is derived from the client adapter and RADIUS server, to encrypt data. Once enabled, a few configuration parameters must be set within the operating system.
  - ***EAP-MD5*** EAP-MD5 is enabled or disabled through the operating system and uses static WEP to encrypt data. EAP-MD5 requires you to enter a separate EAP username and password (in addition to your standard Windows network login) in order to start the EAP authentication process and gain access to the network.

When you enable Network EAP on your access point and configure your client adapter for LEAP, EAP-TLS, or EAP-MD5, or enable Require EAP on your access point and configure your client adapter for EAP-TLS or EAP-MD5 using Windows XP, authentication to the network occurs in the following sequence:

1. The client associates to an access point and begins the authentication process. The client does not gain access to the network until mutual authentication between the client and the RADIUS server is successful.

2. Communicating through the access point, the client and RADIUS server complete a mutual authentication process, with the password (or certificate for EAP-TLS) being the shared secret for authentication. The password (or certificate) is never transmitted

during the process. The authentication process is now complete for EAP-MD5. For LEAP or EAP-TLS, the process continues.

3. If mutual authentication is successful, the client and RADIUS server derive a dynamic, session-based WEP key that is unique to the client.

4. The RADIUS server transmits the key to the access point using a secure channel on the wired LAN.

5. For the length of a session, or time period, the access point and the client use this key to encrypt or decrypt all unicast packets that travel between them.

Using Static WEP

Entering a New Static WEP Key   Using WEP provides a level of security for the network. The following guidelines explain and expand on the settings. Using WEP on the wireless network adds an additional level of security.

1. Select None from the Network Security Type drop-down box on the Network Security screen.

2. Select Use Static WEP Keys under WEP.

3. Select one of the following WEP key entry methods:

    - **Hexadecimal (0-9, A-F)**   Specifies that the WEP key will be entered in hexadecimal characters, which include 0-9, A-F, and a-f.

    - **ASCII text**   Specifies that the WEP key will be entered in ASCII text, which includes alpha characters, numbers, and punctuation marks.

4. Select one of the following access point authentication options, which defines how your client adapter will attempt to authenticate to an access point:

    - **Open authentication**   Allows your client adapter, regardless of its WEP settings, to authenticate and attempt to communicate with an access point. Open Authentication is the default setting.

- **Shared Key Authentication**   Allows your client adapter to communicate only with access points that have the same WEP key. This option is available only if Use Static WEP Keys is selected.

  In shared key authentication, the access point sends a known unencrypted "challenge packet" to the client adapter, which encrypts the packet and sends it back to the access point. The access point attempts to decrypt the encrypted packet and sends an authentication response packet indicating the success or failure of the decryption back to the client adapter. If the packet is successfully encrypted/decrypted, the user is considered to be authenticated.

5. For the static WEP key that you are entering (1, 2, 3, or 4), select a WEP key size of 40 or 128 on the right side of the screen. 128-bit client adapters can use 40- or 128-bit keys, but 40-bit adapters can use only 40-bit keys. If 128 bit is not supported by the client adapter, this option is grayed out, and you are unable to select it.

6. Obtain the static WEP key from your system administrator and enter it in the blank field for the key you are creating. Follow these guidelines to enter a new static WEP key:

   - WEP keys must contain the following number of characters:
     - 10 hexadecimal characters or 5 ASCII text characters for 40-bit keys

       Example: 5A5A313859 (hexadecimal) or ZZ18Y (ASCII)

     - 26 hexadecimal characters or 13 ASCII text characters for 128-bit keys

       Example: 5A5831353335554595549333534 (hexadecimal) or ZX1535TYUI354 (ASCII)

   - Your client adapter's WEP key must match the WEP key used by the access point (in infrastructure mode) or clients (in ad hoc mode) with which you are planning to communicate.

Wireless Networks and Windows XP

- When setting more than one WEP key, the keys must be assigned to the same WEP key numbers for all devices. For example, WEP key 2 must be WEP key number 2 on all devices. When multiple WEP keys are set, they must be in the same order on all devices.

7. Click the Transmit Key button to the left of the key you want to use to transmit packets. Only one WEP key can be selected as the transmit key.
8. Click OK to return to the Profile Manager screen; then click OK or Apply to save your changes.

Overwriting an Existing Static WEP Key

Follow these steps to overwrite an existing static WEP key.

NOTE: *You can overwrite existing WEP keys, but you cannot edit or delete them.*

1. Look at the current WEP key settings in the middle of the Network Security screen. For security reasons, the codes for existing static WEP keys do not appear on the screen.
2. Decide which existing static WEP key you want to overwrite.
3. Click within the blank field of that key.
4. Enter a new key, following the guidelines outlined.
5. Make sure the Transmit Key button to the left of your key is selected, if you want this key to be used to transmit packets.
6. Click OK to return to the Profile Manager screen; then click OK or Apply to save your changes.

Disabling Static WEP

If you ever need to disable static WEP for a particular profile, select No WEP under WEP on the Network Security screen, click OK, and then click OK or Apply on the Profile Manager screen.

Performing Diagnostics

This section explains how to use ACU to perform user-level diagnostics.

Viewing the Current Status of Your Client Adapter

The adapter utility enables you to view the current status of your client adapter as well as many of the settings that have been configured for the adapter.

To view your client adapter's status and settings, open the adapter utility and then click the Status icon or select Status from the Commands drop-down menu.

Viewing Statistics for Your Client Adapter

The adapter utility enables you to view statistics that indicate how data is being received and transmitted by your client adapter.

To view your client adapter's statistics, open ACU; then click the Statistics icon or select Statistics from the Commands drop-down menu.

NOTE: *The receive and transmit statistics are host statistics. That is, they show packets and errors received or sent by the Windows device. Link status tests from the access point or site survey tool are performed at the firmware level; therefore, they have no effect on the statistics shown in the Statistics screen.*

The statistics are calculated as soon as your client adapter is started or the Reset button is selected, and they are continually updated at the rate specified by the screen update timer.

Viewing the Link Status Meter

The adapter's utility link status meter can be used to assess the performance of your client adapter's RF link. If this tool is used to assess the RF link at various locations, you can avoid areas where performance is weak and eliminate the risk of losing the connection between your client adapter and an access point.

To open the link status meter, open the adapter utility; then click the Link Status Meter icon or select Link Status Meter from the Commands drop-down menu.

The Link Status Meter screen provides a graphical display of the following:

- ***Signal strength*** The strength of the client adapter's radio signal at the time packets are being received. It is displayed as a percentage along the vertical axis.

- ***Signal quality*** The quality of the client adapter's radio signal at the time packets are being received. It is displayed as a percentage along the horizontal axis.

The combined result of the signal strength and signal quality is represented by a diagonal line. Where the line falls on the graphical display determines whether the RF link between your client adapter and its associated access point is poor, fair, good, or excellent. The access point that is associated to your client adapter and its MAC address are indicated at the bottom of the display.

If you want to see a recent history of the RF performance between your client adapter and its associated access point, select the Show History checkbox on the Client Utility Preferences screen. Black dots on the graphical display show the performance of the last 50 signals.

Running an RF Link Test

The adapter's utility link test tool sends out pings to assess the performance of the RF link. The test is designed to be performed multiple times at various locations throughout your area and is run at the data rate set on ACU's RF Network Properties screen. The results of the link test can be used to determine RF network coverage and ultimately the required number and placement of access points on your network. The test also helps you to avoid areas where performance is weak, thereby eliminating the risk of losing the connection between your client adapter and its associated access point.

Because the link test operates above the RF level, it does more than test the RF link between two network devices. It also checks the status of wired sections of the network and verifies that TCP/IP and the proper drivers have been loaded. The following section outlines the prerequisites that are required before you can run an RF link test.

Routine Procedures

Once a network card is inserted, generally, there is no need to remove it or to make any changes. However, if the need arises, the following steps outline some things to consider when performing some routing procesures.

Inserting and Removing a Client Adapter

To insert or remove the client adapter, follow the instructions for the device. In most cases, following the procedure specified by the manufacturer is the best practice.

Upgrading the Firmware The client adapter is shipped with the firmware installed in its flash memory; however, a more recent version of the firmware may be available from the manufacturer's web site. Best practice recommends using the most current version of radio firmware after it has been tested on a few machines.

Determining the Firmware Version Look at the available filenames for radio firmware provided by the vendor and verify that the firmware version of the new firmware is newer than the current version. If the firmware available from the vendor's web site is newer than the firmware currently installed in your client adapter, follow the instructions provided by the vendor to upgrade the firmware.

Loading New Firmware Download the latest driver from the Web and install it using the client adapter instructions. Open client utility and then click the Load Firmware icon or select Load New Firmware from the Commands drop-down menu.

Driver Procedures

This section includes the following procedures.

Determining the Driver Version To determine the version of the driver that your client adapter is currently using, open the client utility; then locate the Status icon or select Status from the options. The Status screen displays the current version of your adapter's driver in the NDIS Driver Version field.

Upgrading the Driver To update the driver, follow the instructions for your specific operating system.

Upgrading the Driver for Windows 2000 To update your Windows 2000 computer, follow the instructions listed:

1. Use your computer's web browser to access the vendor's web site.
2. Select the latest driver file for Windows 2000.
3. Save the file to a floppy disk or to your computer's hard drive and unzip it.
4. Make sure your client adapter is installed on your computer.
5. Double-click My Computer, Control Panel, and System.
6. Click the Hardware tab and Device Manager.

7. Double-click Network Adapters and the Cisco Systems wireless LAN adapter.
8. Click the Driver tab.
9. Click the Update Driver button.
10. The Update Device Driver Wizard window appears. Click Next.
11. Select "Display a list of the known drivers for this device so that I can choose a specific driver" and click Next.
12. Click Have Disk.
13. Enter or browse to the path of the new driver and click OK.
14. A message appears, indicating that the system is ready to install the new driver. Click Next and Finish.

The driver upgrade is complete, and the old driver is overwritten by the new one.

Upgrading the Driver for Windows XP   These instructions for upgrading the driver assume you are using Windows XP's classic view. To switch from category view to classic view, double-click Control Panel and then select Switch to Classic View. Then follow these steps:

1. Use your computer's web browser to access the vendor's web site.
2. Select the latest driver file for Windows XP.
3. Read and accept the terms and conditions of the software license agreement.
4. Save the file to a floppy disk or to your computer's hard drive and unzip it.
5. Make sure your client adapter is installed on your computer.
6. Double-click My Computer, Control Panel, and System.
7. Click the Hardware tab and Device Manager.
8. Double-click Network Adapters and Wireless LAN Adapter.
9. Click the Driver tab and the Update Driver button. The Welcome to the Hardware Update Wizard screen appears.

10. Select the Install from a list or specific location (Advanced) option and click Next.
11. When prompted to choose your search and installation options, select "Don't search. I will choose the driver to install" and click Next.
12. When prompted to select a network adapter to install, click the Have Disk button. The Install From Disk screen appears.
13. Click the Browse button, find the location of the driver (on your floppy disk or computer's hard drive), and click Open. The installation wizard finds the driver file. Click OK on the Install From Disk screen.
14. The Select Network Adapter screen reappears. Select the driver for your wireless LAN adapter and click Next.
15. The installation wizard copies the driver files from the floppy disk or computer's hard drive. When the installation is complete, click Finish.

The driver upgrade is complete, and the old driver is overwritten by the new one.

Troubleshooting

This section provides information for diagnosing and correcting common problems encountered when installing or operating WLAN adapters.

Client Adapter Recognition Problems

If your computer's PCMCIA adapter does not recognize your client adapter, check your computer's *basic input/output system* (BIOS) and make sure that the PC card controller mode is set to *Peripheral Component Interconnect Core* (PCIC) compatible.

Resolving Resource Conflicts

If you encounter problems while installing your client adapter on a computer running a Windows operating system, you may need to specify a different *interrupt request* (IRQ) or I/O range for the adapter.

The default IRQ for the client adapter may not work for all systems. Follow the steps for your specific operating system to obtain an available IRQ.

During installation, the adapter's driver installation script scans for an unused I/O range. The installation can fail if the I/O range found by the driver installation script is occupied by another device but not reported by Windows. An I/O range might not be reported if a device is physically present in the system but not enabled under Windows. Follow the steps for your specific operating system to obtain an available I/O range.

Resolving Resource Conflicts in Windows 2000 In case a conflict occurs with Windows 2000, follow the steps listed:

1. Double-click My Computer, Control Panel, and System.
2. Click the Hardware tab and Device Manager.
3. Double-click Network Adapters and the Cisco Systems wireless LAN adapter.
4. In the General screen, the Device Status field indicates if a resource problem exists. If a problem is indicated, click the Resources tab.
5. Deselect the Use automatic settings checkbox.
6. Under Resource Settings or Resource Type, click Input/Output Range.
7. Look in the Conflicting Device list at the bottom of the screen. If it indicates that the range is being used by another device, click the Change Setting button.
8. Scroll through the ranges in the Value dialog box and select one that does not conflict with another device. The Conflict

Wireless Networks and Windows XP

Information window at the bottom of the screen indicates if the range is already being used.

9. Click OK.
10. Under Resource Settings or Resource Type, click Interrupt Request.
11. Look in the Conflicting Device list at the bottom of the screen. If it indicates that the IRQ is being used by another device, click the Change Setting button.
12. Scroll through the IRQs in the Value dialog box and select one that does not conflict with another device. The Conflict Information window at the bottom of the screen indicates if the IRQ is already being used.
13. Click OK.
14. Reboot your computer.

Resolving Resource Conflicts in Windows XP To resolve a conflict with a Windows XP based computer, follow the instructions listed:

1. Double-click My Computer, Control Panel, and System.
2. Click the Hardware tab and Device Manager.
3. Under Network Adapters, double-click Wireless LAN Adapter.
4. In the General screen, the Device Status field indicates if a resource problem exists. If a problem is indicated, click the Resources tab.
5. Deselect the Use automatic settings checkbox.
6. Under Resource Settings, click I/O Range.
7. Look in the Conflicting Device list at the bottom of the screen. If it indicates that the range is being used by another device, click the Change Setting button.
8. Scroll through the ranges in the Value dialog box and select one that does not conflict with another device. The Conflict Information window at the bottom of the screen indicates if the range is already being used.

9. Click OK.
10. Under Resource Settings, click IRQ.
11. Look in the Conflicting Device list at the bottom of the screen. If it indicates that the IRQ is being used by another device, click the Change Setting button.
12. Scroll through the IRQs in the Value dialog box and select one that does not conflict with another device. The Conflict Information window at the bottom of the screen indicates if the IRQ is already being used.
13. Click OK.
14. Reboot your computer.

Problems Associating with an Access Point

Follow these instructions if your client adapter fails to associate with an access point:

1. If possible, move your workstation a few feet closer to an access point and try again.
2. Make sure the client adapter is securely inserted in your computer's client adapter slot.
3. If you are using a PCI client adapter, make sure the antenna is securely attached.
4. Make sure the access point is turned on and operating.
5. Check that all parameters are set properly for both the client adapter and the access point. These include the SSID, EAP authentication, WEP activation, network type, channel, and so on.
6. Follow the instructions in the previous section to resolve any resource conflicts. If you are using Windows NT, you may also want to try disabling the Ethernet port.
7. If the client adapter still fails to establish contact, refer to the section in the Preface for technical support information.

Problems Authenticating with an Access Point

If your client adapter is a 40-bit card and LEAP or EAP is enabled, the adapter can associate to but not authenticate to access points using 128-bit encryption. To authenticate to an access point using 128-bit encryption, you have two options:

- Purchase a 128-bit client adapter. This is the most secure option.
- Disable static WEP for the client adapter and configure the adapter and the access point to associate with mixed cells. This option presents a security risk because your data is not encrypted as it is sent over the RF network.

Problems Connecting to the Network

If you continue to experience problems after you have installed the appropriate driver and client utilities, check the Proxy server, network protocols, and authentication information.

Configuring the Client Adapter Through Windows XP

This section provides instructions for minimally configuring the client adapter through Windows XP (instead of through ACU) as well as for enabling one of the three security options that are available for use with this operating system. In addition, this section also provides basic information on using Windows XP to specify the networks to which the client adapter associates and to view the current status of your client adapter. If you require more information about configuring or using your client adapter with Windows XP, refer to Microsoft's documentation for Windows XP.

Overview of Security Features

Setting the WEP keys is a good practice for deploying a wireless network. The following sections provide a better understanding of the types of WEP security available.

Static WEP Keys

Each device within the wireless network can be assigned up to four static WEP keys. If a device receives a packet that is not encrypted with the appropriate key (as the WEP keys of all devices that are to communicate with each other must match), the device discards the packet and never delivers it to the intended receiver.

Static WEP keys are write-only and temporary; therefore, they cannot be read back from the client adapter, and they are lost when power to the adapter is removed or when the Windows device is rebooted. Although the keys are temporary, you do not need to re-enter them each time the client adapter is inserted or the Windows device is rebooted. This is because the keys are stored (in an encrypted format for security reasons) in the registry of the Windows device. When the driver loads and reads the client adapter's registry parameters, it also finds the static WEP keys, unencrypts them, and stores them in volatile memory on the adapter.

EAP (with Static or Dynamic WEP Keys)

The new standard for WLAN security, as defined by the IEEE, is called 802.1X for 802.11, or simply 802.1X. An access point that supports 802.1X and its protocol, EAP, acts as the interface between a wireless client and an authentication server, such as a RADIUS server, to which the access point communicates over the wired network.

Two 802.1X authentication types are available when configuring your client adapter through Windows XP:

- ***EAP-TLS*** This authentication type is enabled through the operating system and uses a dynamic, session-based WEP key, which is derived from the client adapter and RADIUS server, to

Wireless Networks and Windows XP

encrypt data. EAP-TLS requires the use of a certificate. Refer to Microsoft's documentation for information on downloading and installing the certificate.

- ***EAP-MD5*** This authentication type is enabled through the operating system and uses static WEP to encrypt data. EAP-MD5 requires you to enter a separate EAP username and password (in addition to your standard Windows network login) in order to start the EAP authentication process and gain access to the network. If you want to authenticate without encrypting the data that is transmitted over your network, you can use EAP-MD5 without static WEP.

When you enable EAP on your access point and configure your client adapter for EAP-TLS or EAP-MD5 using Windows XP, authentication to the network occurs in the following sequence:

1. The client adapter associates with an access point and begins the authentication process.
2. Communicating through the access point, the client, and RADIUS server complete a mutual authentication process, with the password (for EAP-MD5) or certificate (for EAP-TLS) being the shared secret for authentication. The password or certificate is never transmitted during the process.
3. If mutual authentication is successful, the client and RADIUS server derive a dynamic, session-based WEP key that is unique to the client.
4. The RADIUS server transmits the key to the access point using a secure channel on the wired LAN.
5. For the length of a session, or time period, the access point and the client use this key to encrypt or decrypt all unicast packets that travel between them.

Configuring the Client Adapter

If you installed the adapter utility but intend to use Windows XP to configure the client adapter, open the adapter utility and make sure the Allow Windows to Configure My Wireless Network Settings option is selected on the Profile Manager screen.

Select the Network Authentication (Shared mode) checkbox if you want to use shared key, rather than open, authentication with the access point.

Open authentication enables your client adapter, regardless of its WEP settings, to authenticate and attempt to communicate with an access point.

Shared key authentication allows your client adapter to communicate only with access points that have the same WEP key. It is not recommended to use shared key authentication because it presents a security risk.

If you are planning to use EAP-TLS authentication, do not select the EAP-TLS checkbox. EAP-TLS does not work with shared key authentication because shared key authentication requires the use of a WEP key, and a WEP key is not set for EAP-TLS until after the completion of EAP authentication. The WEP key must be assigned to the same number on both the client adapter and the access point (in an infrastructure network) or other clients (in an ad hoc network).

Using Windows XP to Associate with an Access Point

Windows XP causes the client adapter's driver to automatically attempt to associate with the first network in the list of preferred networks. If the adapter fails to associate or loses association, it automatically switches to the next network in the list of preferred networks. The adapter does not switch networks as long as it remains associated to the access point. To force the client adapter to associate with a different access point, you must select a different network from the list of available networks (and click Configure and OK).

Viewing the Current Status of Your Client Adapter

To view the status of your client adapter, click the icon of the two connected computers in the Windows system tray. The Wireless Network Connection Status screen appears (see Figure 5-2).

Figure 5-2
Wireless Network Connection Status screen

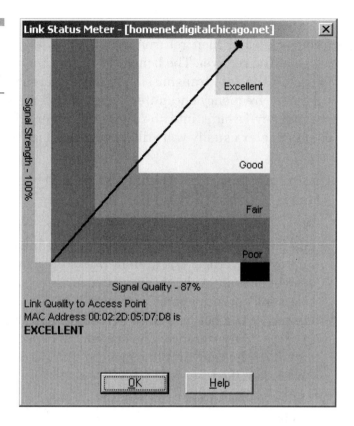

Installation and Configuration Guide for Windows CE

Hardware Components

The client adapter has three major hardware components: a radio, a radio antenna, and LEDs.

Radio The client adapter contains a *direct sequence spread spectrum* (DSSS) radio that operates in the 2.4 GHz license-free *Industrial Scientific Medical* (ISM) band. The radio transmits data over a half-duplex radio channel operating at up to 11 Mbps.

DSSS technology distributes a radio signal over a wide range of frequencies and then returns the signal to the original frequency range at the receiver. The benefit of this technology is its capability to protect the data transmission from interference. For example, if a particular frequency encounters noise, interference, or both, enough redundancy is built into the signal on other frequencies that the client adapter usually will still be successful in its transmission.

Radio Antenna The type of antenna used depends on your client adapter:

- *PC cards* have an integrated, permanently attached diversity antenna. The benefit of the diversity antenna system is improved coverage. The system works by allowing the card to switch and sample between its two antenna ports in order to select the optimum port for receiving data packets. As a result, the card has a better chance of maintaining the RF connection in areas of interference. The antenna is housed within the section of the card that hangs out of the PC card slot when the card is installed.
- *LM cards* are shipped without an antenna; however, an antenna can be connected through the card's external connector.

External antennas used in combination with a power setting resulting in a radiated power level above 100 mW *equivalent isotropic radiated power* (EIRP) are not allowed for use within the European community and other countries that have adopted the European *Radio and Telecommunuications Terminal Equipment* (R&TTE) directive or the CEPT recommendation, Rec 70.03, or both.

LEDs The client adapter has two LEDs that glow or blink to indicate the status of the adapter or to convey error messages.

Software Components

The client adapter has three major software components: radio firmware, a driver, and client utilities.

Radio Firmware The firmware, which is contained in the client adapter's flash memory, controls the adapter's radio. The client adapter is shipped with the firmware installed; however, a more recent version of the firmware may be available from Cisco.com.

Driver The driver provides an interface between the Windows CE device and the client adapter, thereby enabling Windows CE and the applications it runs to communicate with the adapter. The driver is provided on the Wireless LAN Adapters CD that shipped with the client adapter and must be installed before the adapter can be used. The CD has the latest version of the driver available at the time of shipping; however, a more recent version of the driver may be available from the vendor's web site.

Client Utilities The client utilities are optional applications that interact with the radio firmware to adjust client adapter settings and display information about the adapter. The client utilities and online help files are installed with the driver. The CD has the latest version of the client utilities, but a more recent version of the client utilities may be available from the vendor's web site.

Network Configurations Using the Client Adapter

The client adapter can be used in a variety of network configurations. In some configurations, access points provide connections to your network or act as repeaters to increase the wireless communication range. The maximum communication range is based on how you configure your wireless network.

This section describes and illustrates the two most common network configurations:

- Ad hoc WLAN
- Wireless infrastructure with workstations accessing a wired LAN

Ad Hoc WLAN An ad hoc (or peer-to-peer) WLAN is the simplest WLAN configuration. In a WLAN using an ad hoc network configuration, all devices equipped with a client adapter can be linked together and communicate directly with each other.

Wireless Infrastructure with Workstations Accessing a Wired LAN A microcellular network can be created by placing two or more access points on a LAN. In a microcellular network, workstations access a wired LAN through one or several access points. This configuration is useful with portable or mobile stations because it enables them to be directly connected to the wired network even while moving from one microcell domain to another (see Figure 5-3). This process is transparent, and the connection to the file server or host is maintained without disruption. The mobile station stays connected to an access point as long as it can. However, once the transfer of data packets needs to be retried or beacons are missed, the station automatically searches for and associates with another access point. This process is referred to as *seamless roaming*.

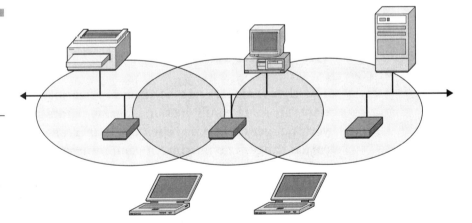

Figure 5-3
Wireless infrastructure with workstations accessing a wired LAN

Preparing for Installation

Topics related to installation in a pocket-PC device are covered in this section.

Unpacking the Client Adapter

Open the shipping container and carefully remove the contents. Return all packing materials to the shipping container and save it. The adapter and CD are the items necessary for installation and configuration.

System Requirements

In addition to the items shipped with the client adapter, you will also need the following in order to install the adapter:

- One of the following Windows CE devices equipped with a Type II or Type III PC card slot:
  - *Handheld PC* (HPC) running Windows CE 2.11 with an ARM, StrongARM, Mips, SH3, SH4, x86, or Pentium processor.
  - HPC running Windows CE 3.0 with an ARM, StrongARM, Mips, x86, or Pentium processor.
  - *Pocket PC* (PPC) running Windows CE 3.0 with an ARM, StrongARM, Mips, or SH3 processor.
  - All drivers and supporting software (card and socket services) for the PC card slot must be loaded and configured.
- Laptop or PC with a CD-ROM drive and running a Windows operating system and ActiveSync

- Serial or *Universal Serial Bus* (USB) connection to the Windows CE device
  - The logical name for your Windows CE device (also referred to as *client name*)
  - The case-sensitive SSID for your RF network
  - The primary and secondary *Domain Name System* (DNS) and *Windows Internet Name Service* (WINS) to be assigned to your Windows CE device
  - If your Windows CE device is not connected to a DHCP server, the IP address, subnet mask, and default gateway address to be assigned to your device
  - The WEP keys of the access points with which your client adapter will communicate, if your wireless network uses static WEP for security
  - The username and password for your RADIUS server account, if your wireless network uses server-based authentication

Installing the Client Adapter

Installation of the client adapter and the associated driver is slightly different for Windows CE devices. It requires using the syncronization software that came with the device.

Installing the Driver and Client Utilities

The installation files for Windows CE devices, which are used to install the driver and client utilities, consist of individual *.cab files. Each *.cab file contains the driver, client utility, and online help files for a specific Windows CE CPU/device combination. When the *.exe file is executed, all the *.cab files are extracted, and the *.cab file specific to the attached Windows CE device is copied to the device.

This procedure is meant to be used the first time the driver and client utilities are installed on a Windows CE device. If the client adapter software is already installed on your Windows CE device,

Wireless Networks and Windows XP

follow the instructions to first uninstall any existing software and then upgrade to new software.

The driver and client utilities must be installed before you insert a client adapter into a Windows CE device:

1. Use a serial or USB cable to connect your Windows CE device to a laptop or PC running ActiveSync. A message appears on the Windows CE device indicating that it is connecting to the host. After the Windows CE device is connected, the New Partnership window appears on the laptop or PC. This window asks if you want to set up a partnership.

2. Perform one of the following:
   - If you want to establish a partnership that allows you to synchronize files between the laptop or PC and the Windows CE device, select Yes, click Next, and follow the instructions on the screen to specify the files to be synchronized and to finish setting up the partnership.
   - If you do not want to synchronize files and want to connect as a guest, select No and click Next. The screen indicates that you are connected as a guest.

3. If you are installing the driver and client utilities from the wireless adapter's CD that shipped with the client adapter, follow these steps:
   a. Insert the CD into the laptop or PC's CD-ROM drive.
   b. Use Windows Explorer to access the WinCE directory.
   c. Select the Win CE 2.1 or Win CE 3.0 directory, depending on which version of Windows CE your device is running.
   d. Select the .exe file.
   e. Read and accept the terms and conditions of the software license agreement.
   f. Save the file to a floppy disk or to the hard drive of your laptop or PC.

4. Double-click the *.exe file for your version of Windows CE. The application creates an Install directory under the ActiveSync directory, extracts the .cab files contained in the *.exe file, and copies them to the Install directory.

5. Click Next to start the Windows CE Application Manager (CeAppMgr), which is installed with ActiveSync. CeAppMgr interrogates the Windows CE device to determine its processor type.

   If a Windows CE device is not connected to the laptop or PC, click Exit to quit the setup program and connect a Windows CE device or click Next to continue the installation. If you select Next, a message appears, indicating that the software will be downloaded the next time a mobile device is connected. Click OK. The next time a Windows CE device is connected to the laptop or PC via ActiveSyn, CeAppMgr starts automatically, and you are prompted to install the software. If you select Exit, click OK to shut down CeAppMgr and start again.

6. When a dialog box appears asking if you want to install the client adapter using the default application installation directory, click Yes. The default directory is \Windows\Programs\ on HPC devices and \Windows\Start Menu\Programs\ on PPC devices.

   If you click No on an HPC device, CeAppMgr transfers the *.cab file to the Windows CE device and executes it. This process takes awhile and shows no evidence of activity. Eventually, a screen appears on the Windows CE device that asks you where the application files should be installed.

   A message and a progress bar appear, indicating that the client adapter is being installed. CeAppMgr copies the processor-specific *.cab file to the Windows CE device. Then the driver and help files are copied to the \Windows directory, and the client utilities are installed in the \Windows\Programs\vendor directory on HPC devices or the \Windows\Start Menu\Programs\vendor directory on PPC devices. Shortcuts to the adapter utility are automatically added to the desktop on HPC devices.

7. When the installation process is complete on the laptop or PC, a message appears asking you to check the screen of the Windows CE device to see if any additional steps are required to complete the installation. Click OK to terminate the installation process on the laptop or PC.

8. Complete any required steps on the Windows CE device.
9. Remove the CD, if installed.
10. Disconnect the Windows CE device.
11. Insert the client adapter into the PC card slot of the Windows CE device.

    The Windows CE device should configure the client adapter, and the green LED on the adapter should blink. If this does not happen, remove the client adapter, reset the Windows CE device, and reinsert the client adapter.

12. The Wireless LAN Adapter Settings dialog box appears. (If the dialog box does not appear, select Start, Settings, Control Panel, Network, the Adapters tab, the Wireless LAN adapter, and Properties on HPC devices or select Start, Settings, the Connections tab, Network, and the Wireless LAN adapter on PPC devices.)

13. Perform one of the following:
    - If your device is connected to a DHCP server, select "Obtain an IP address via DHCP" or "Use server-assigned IP address" and click OK.
    - If your device is not connected to a DHCP server, select "Specify an IP address" or "Use specific IP address" and follow these steps:
        a. Enter the IP address, subnet mask, and default gateway address you want to assign to your device. They can be obtained from your system administrator.
        b. Select the Name Servers tab and enter the primary and secondary DNS and WINS you want to assign to your device. They can be obtained from your system administrator.
        c. Click OK.

14. Double-click the adapter utility icon to open ACU.
15. Select SSID under Property. Then enter your RF network's case-sensitive SSID in the Value box. The SSID identifies the specific wireless network that you want to access. The range is up to 32 characters and is case sensitive.

If you leave this parameter blank, your client adapter can associate with any access point on the network that is configured to allow broadcast SSIDs (see the AP Radio Hardware page in the access point management system). If the access point with which the client adapter is to communicate is not configured to allow broadcast SSIDs, the value of this parameter must match the SSID of the access point. Otherwise, the client adapter cannot access the network.

16. Select Client Name under Property. Then enter your Windows CE device's unique client name in the Value box.

    A client name is a logical name for your Windows CE device. It allows an administrator to determine which devices are connected to the access point without having to memorize every MAC address. This name is included in the access point's list of connected devices. The range here is up to 16 characters. Each device on the network should have a unique client name.

17. Select Data Rates under Property. Make sure that Auto is selected in the list of options in the Value box.

    The Data Rates setting specifies the rate at which you want your client adapter to transmit or receive packets to or from access points (in infrastructure mode) or other clients (in ad hoc mode). Auto is recommended for infrastructure mode; setting a specific data rate is recommended for ad hoc mode.

    The client adapter's data rate must be set to Auto or must match the data rate of the access point (in infrastructure mode) or the other clients (in ad hoc mode) with which it is to communicate. Otherwise, your client adapter may not be able to associate with them.

18. Click OK. The driver and client utility installation is complete. The client adapter has been installed and configured for basic operation.

Verifying Installation

To verify that you have properly installed the driver and client utilities and minimally configured your client adapter, check the client adapter's LEDs. If the installation was successful, the client adapter's LED blinks.

If your installation was unsuccessful or you experienced problems during or after driver installation, refer to the troubleshooting information.

Enabling Security Features

In order for the Windows CE device to communicate with the wireless network or other wireless devices, the correct SSID and WEP settings must be configured to match the network settings. Use the following guidelines.

Using Static WEP

To enter new, static WEP keys or overwrite existing static WEP keys, follow these steps for the vendor's implementation:

1. View the client adapter's current static WEP key settings.
2. Perform one of the following:
    a. Enter a new static WEP key and enable WEP.
    b. Overwrite an existing static WEP key.

Entering a New Static WEP Key and Enabling Static WEP

Follow these steps to enter a new static WEP key for your client adapter:

1. If you entered the password correctly in the Enter Password screen, the CEM screen appears (see Figure 5-4).

    This screen enables you to create up to four static WEP keys.

2. For the static WEP key that you are entering (1, 2, 3, or 4), select a WEP key size of 40 or 128 on the right side of the screen. 128-bit client adapters can use 40- or 128-bit keys, but 40-bit

Figure 5-4
CEM screen

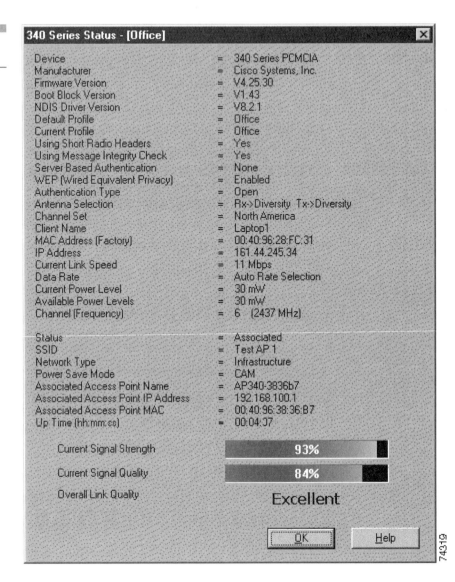

adapters can use only 40-bit keys. If 128 bit is not supported by the client adapter, this option is grayed out, and you are unable to select it.

3. Obtain the static WEP key from your system administrator and enter it in the blank field for the key you are creating. Follow these guidelines to enter a new static WEP key:

Wireless Networks and Windows XP

- WEP keys can consist of the following hexadecimal characters: 0-9, A-F, and a-f.
- WEP keys must contain the following number of characters:
  - 10 hexadecimal characters for 40-bit keys
    Example: 12345abcde
  - 26 hexadecimal characters for 128-bit keys
    Example: AB34CD78EFab01cd23ef456789
- Your client adapter's WEP key must match the WEP key used by the access point (in infrastructure mode) or clients (in ad hoc mode) with which you are planning to communicate.
- When setting more than one WEP key, the keys must be assigned to the same WEP key numbers for all devices. For example, WEP key 2 must be WEP key number 2 on all devices. When multiple WEP keys are set, they must be in the same order on all devices.

NOTE: *After you enter a WEP key, you can write over it, but you cannot edit or delete it.*

4. Click the Transmit Key button to the left of the key you want to use to transmit packets. Only one WEP key can be selected as the transmit key.

5. Click OK to write your WEP key(s) to the client adapter's volatile memory and the Windows CE registry and to exit the utility. Or click Cancel to exit the utility without updating the keys.

6. Double-click the adapter utility icon. The adapter utility screen appears.

7. Select WEP under Property and then select Enabled from the list of options in the Value box to enable static WEP.

8. Click OK to save your changes and to exit the utility.

Overwriting an Existing Static WEP Key

Follow these steps to overwrite an existing static WEP key:

1. If you entered the password correctly in the Enter Password screen, the CEM screen appears.

   A checkmark appears in the Already Set? box for all existing static WEP keys.
2. Decide which existing static WEP key you want to overwrite.
3. Click within the blank field of that key.
4. Enter a new key, following the currently existing guidelines.
5. Make sure the Transmit Key button to the left of your key is selected if you want this key to be used to transmit packets.
6. Click OK to write your new static WEP key to the client adapter's volatile memory and the Windows CE registry and to exit the utility. Or click Cancel to exit the utility without overwriting any keys.
7. Open the adapter utility and make sure WEP is enabled.

Disabling Static WEP

Follow these steps if you ever need to disable static WEP:

1. Double-click the adapter utility icon and the adapter utility screen appears.
2. Select WEP under Property and then select Disabled from the list of options in the Value box to disable WEP.
3. Click OK to save your changes and to exit the utility.

Windows CE-based Device Notes

If users have a Pocket PC device running Windows CE, then they have the ability to easily connect to the wireless corporate network and to other wireless networks. Microsoft, Cisco, Lucent, Compaq,

Hewlett-Packard, and other vendors provide complete support for connecting the Pocket PC to the wireless network. The devices are ready for wireless networking with software drivers and applications, making them easy to deploy and useful in the wireless enterprise.

To use a Compaq iPaq with a wireless network, the iPaq will need a PC Card sleeve and an iPaq-compatible wireless network card. Compaq and Lucent are just two of the many vendors that produce a wireless network card compatible with the Compaq iPaq.

In the enterprise, the Pocket PC, in this case a Compaq iPaq, will connect to the access point much the same way as other laptop and desktop computers. It is possible to connect the Pocket PC device with a laptop in ad hoc mode, but with an infrastructure network already in place, connectivity is greatly simplified.

Some manufacturers are selling HomeRF-based wireless network cards for the Pocket PC device. HomeRF is another wireless networking standard for the home, and HomeRF network cards are not compatible with the 802.11 standard.

Using WEP with a Windows CE-based Device

If the wireless network is running WEP, then security settings for the Pocket PC need to be entered in the device. The same SSID and WEP settings used for computers will need to be used for the Pocket PC-based device. Software settings are individual to the version of Windows CE and the wireless network card (wireless NIC). Consult the documentation for the settings and where to configure the settings. Most vendors provide excellent documentation for connecting the Pocket PC device with a WLAN.

Using DHCP with a Windows CE-based Device

Windows CE-based devices can communicate with a DHCP server for IP address assignment. Both devices rely on industry standards to request and receive an IP address for communication. Settings for the IP address can be found by clicking the Network icon in the

Connections folder of the Pocket PC device. Specific instructions can be found in the Pocket PC documentation.

Once a Pocket PC device is configured and connected to the network, it can be used to perform site surveys. Carrying around a Pocket PC device is much easier than carrying a laptop that provides the same data.

Advanced Configuration

This section explains how to set the client adapter's advanced configuration parameters.

Configuring Your Client Adapter

The adapter utility enables you to change the configuration parameters of your client adapter. Follow the steps here to open ACU and make any configuration changes.

The driver and client utility installation process provides instructions to initially configure your client adapter and make it operational. Therefore, the information in this section is necessary only if you need to configure your client adapter to utilize an advanced feature.

Performing Diagnostics

This section explains how to use the client utilities to perform user-level diagnostics.

Overview of the Diagnostic Utilities

The diagnostic utilities enable you to assess the performance of your client adapter within the wireless network. These utilities perform the following functions:

Wireless Networks and Windows XP

- Display your client adapter's current status.
- Display statistics pertaining to your client adapter's transmission and reception of data.

Viewing the Current Status of Your Client Adapter

The adapter utility enables you to view the current status of your client adapter (see Figure 5-5).

The first line of the Link Status screen indicates the operational mode of your client adapter and the name or MAC address of any associated access point. The value here can be Associated, Not Associated, Authenticated, or Ad Hoc Mode. The signal strength for all

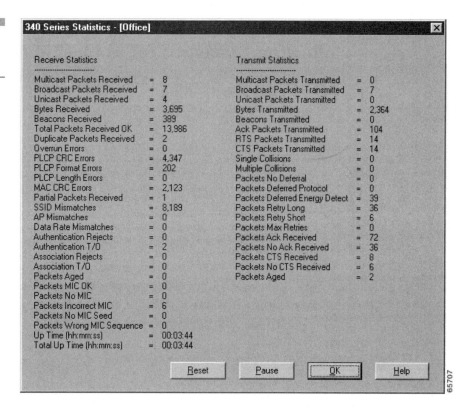

Figure 5-5
Client adapter status

received packets is included here as well. The higher its value and the more green the bar graph is, the stronger the signal. The range can be from 0 to 100 percent.

The signal quality for all received packets is also included here. The higher the value and the more green the bar graph is, the clearer the signal is. Beacons Received is displayed instead if your firmware version is less than 4.05.

The histogram below the bar graph provides a visual interpretation of the current signal quality. Differences in signal quality are indicated by the following colors: green (highest quality), yellow (average), and red (lowest quality). The range here is from 0 to 100 percent.

Overall Link Quality The client adapter's capability to communicate with the access point is determined by the combined result of the adapter's signal strength and signal quality. The value here is Not Associated, Poor, Fair, Good, and Excellent.

Viewing Statistics for Your Client Adapter

The *Client Statistics Utility* (CSU) enables you to view statistics that indicate how data is being received and transmitted by your client adapter.

NOTE: *The receive and transmit statistics are host statistics (see Figure 5-6 and Table 5-2). That is, they show packets and errors received or sent by the Windows CE device. Link status tests from the access point or SST are performed at the firmware level; therefore, they have no effect on the statistics shown by this utility.*

To view your client adapter's statistics, select Start, Programs, Cisco, Client Statistics Utility. The statistics are calculated as soon as your client adapter is started.

Wireless Networks and Windows XP

Figure 5-6
Receive statistics screen

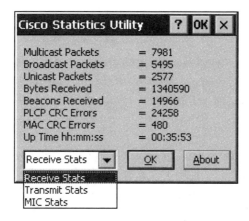

Table 5-2
Receive Statistics

Receive Statistic	Definition
Multicast packets	The number of multicast packets that were received successfully.
Broadcast packets	The number of broadcast packets that were received successfully.
Unicast packets	The number of unicast packets that were received successfully.
Bytes received	The number of bytes of data that were received successfully.
Beacons received	The number of beacon packets that were received successfully.
PLCP CRC errors	The number of times the client adapter started to receive an 802.11 *physical layer convergence protocol* (PLCP) header, but the rest of the packet was ignored due to a *cyclic redundancy check* (CRC) error in the header.

NOTE: CRC errors can be attributed to packet collisions caused by a dense population of client adapters, overlapping access point coverage on a channel, high multipath conditions due to bounced signals, or the presence of other 2.4 GHz signals from devices such as microwave ovens, wireless handset phones, and so on.

MAC CRC Errors These errors occur when a number of packets have a valid 802.11 PLCP header but contain a CRC error in the data portion of the packet.

Up Time (hh:mm:ss) This is the amount of time (in hours:minutes:seconds) since your client adapter was started. If the client adapter has been running for more than 24 hours, the time is displayed in days, hours:minutes:seconds. Figure 5-7 shows an example of a typical configuration. For more information about the headings, see Table 5-3.

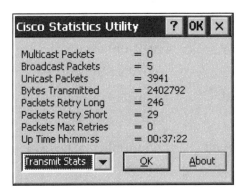

Figure 5-7
Transmit statistics screen

Wireless Networks and Windows XP

Table 5-3
Transmit Statistics

Transmit Statistic	Definition
Multicast packets	The number of multicast packets that were transmitted successfully.
Broadcast packets	The number of broadcast packets that were transmitted successfully.
Unicast packets	The number of unicast packets that were transmitted successfully.
Bytes transmitted	The number of bytes of data that were transmitted successfully.
Packets retry long	The number of normal data packets that were retransmitted.
Packets retry short	The number of RTS packets that were retransmitted.
Packets max retries	The number of packets that failed to be transmitted successfully after exhausting the maximum number of retries.
Up time (hh:mm:ss)	The amount of time (in hours:minutes:seconds) since your client adapter was started. If the client adapter has been running for more than 24 hours, the time is displayed in days, hours:minutes:seconds.

Routine Procedures

This section provides procedures for common tasks related to the client adapter.

Inserting and Removing a PC Card

This section provides instructions for inserting a PC card into or removing a PC card from a Windows CE device.

Inserting a PC Card into a Windows CE Device   Follow these steps to insert a PC card into a Windows CE device:

1. Before you begin, examine the PC card. One end has a dual-row, 68-pin PC card connector. The card is keyed so it can be inserted only one way into the PC card slot.
2. Hold the PC card with the Cisco logo facing up and insert it into the PC card slot, applying just enough pressure to make sure it is fully seated (see Figure 5-8).

Removing a PC Card from a Windows CE Device   To remove a PC card after it is successfully installed and configured, press the Eject button and pull the card out of the PC card slot. When the PC card is reinserted, your connection to the network should be re-established.

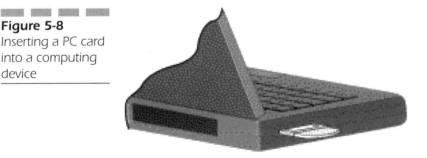

Figure 5-8
Inserting a PC card into a computing device

Upgrading the Client Adapter Software

The following section describes the configuration for a Windows CE device.

Upgrading the Firmware   The client adapter is shipped with the firmware installed in its flash memory; however, a more recent version of the firmware may be available from Cisco.com. Cisco recommends using the most current version of radio firmware. Follow the instructions in this section to determine the version of your client adapter's firmware and to upgrade it if a more recent version is available from Cisco.com.

Upgrading the Driver and Client Utilities   Follow the instructions in this section to determine the versions of your client adapter's driver and client utilities and to upgrade them if more recent versions are available from Cisco.com. The driver, client utilities, and online help files are installed together.

Troubleshooting the Client Adapter

This section provides troubleshooting tips if you encounter problems with your client adapter.

Problems Obtaining an IP Address

If your network is set up to use DHCP to acquire an IP address, the DHCP lease renewal may fail, especially in suspend/resume situations. To force DHCP to try to reacquire an IP address, power your Windows CE device off and on, or eject and reinsert your client adapter.

Problems Connecting to the Network

After you have installed the appropriate driver and client utilities, contact your IS department if you have a problem connecting to the network. Proxy server, DNS or WINS, and further authentication information might be needed to connect to the network.

Error Messages

This section provides a list of error messages that may appear during the installation, configuration, or use of your client adapter. The error messages are listed in alphabetical order, and an explanation as well as a recommended user action are provided for each message.

Getting Help

To access online help for the adapter utility, follow these instructions for your specific Windows CE device.

On HPC Devices

To access help related to the adapter utility on an HPC device, open the utility and click the ? button on the top of the screen. Select the topic for which you want information.

On PPC Devices

To access help related to adapter utility on a PPC device, open the adapter utility and select Start, Help. Select the topic for which you want information.

 **Summary**

WLAN cards work on several devices and OS platforms. Once installed and properly configured, the network cards provide secure, reliable access to a WLAN and beyond. Manufacturers have made it easy to install and configure these devices by including utilities that help configure the cards.

CHAPTER 6

Wireless Home Network Configurations

Deployment Considerations

The best solution is one that meets the needs of your home network at or below the budget. If your company has a wireless network at the office, or if you connect regularly to a corporate wireless network, be sure to consider those requirements and use them as a reference for the home network. Three distinct options are available, each offering various advantages and disadvantages.

Bringing the Wireless Enterprise Home

A number of products on the market are 802.11 compliant. Hardware that is not 802.11 compliant will probably use a proprietary solution and will not work well with products from other vendors. Just like the enterprise, two network architectures, ad hoc and infrastructure can be used for the home.

The ad hoc network is basically like the conference room scenario, multiple computers in the same room just needing to share files and possibly a printer. They are only in need of a peer-to-peer network, and only talk to each other. In this scenario, an access point is not necessary. In a *wireless local area network* (WLAN), an access point enables users to attach to the network just like any other node and share resources such as printers, and broadband Internet connections.

Various WLANs for the Home

There are three basic types of networks you will have the option to set up:

- Ad hoc mode wireless home LAN using Windows *Internet connection sharing* (ICS) with client adapters. This configuration requires one computer running windows and serving as the ICS server.

- Infrastructure Mode wireless Home LAN using Windows ICS with a bridging access point, such as the Buffalo Technologies Airstation with client adapters. In this configuration, all computers, including those serving as servers and workstations, connect through the access point. The Windows ICS server serves as the gateway and provides connectivity to the Internet, if a connection is available.
- Infrastructure Mode wireless Home LAN using a hardware-based *Network Address Translator* (NAT), such as the Buffalo Technologies router access point. In this scenario, the access point also serves as a router and provides additional functionality such as allowing several computers connectivity and access to the Internet.

Ad Hoc Mode Wireless Home LAN Using Windows ICS

The ad hoc approach requires a WLAN card in the Windows ICS system and in the other devices that connect through the Windows ICS system. In addition, this method requires at least two connections, one connection to the Internet (dial-up or broadband) and a second WLAN connection via the wireless network card installed in the ICS computer. This configuration allows for Internet access by wireless workstations (see Figure 6-1) by using the Windows ICS server as a gateway. The ICS server must remain on to provide workstations connectivity with each other and access to the Internet. ICS is a network technology feature included with Windows 2000 and Windows XP.

Infrastructure Mode Wireless Home LAN Using Windows ICS

The infrastructure setup does not require a WLAN card in the Windows ICS system. It does require the use of a wireless access point(s), such as the Buffalo Technologies WLA-L11 bridge/hub access point, many access point products provide the same functionality as the WLA-L11. For an example of an infrastructure see Figure 6-2.

Figure 6-1
Ad hoc mode home LAN topology

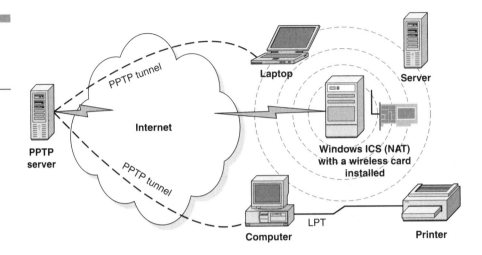

The infrastructure method provides higher flexibility, but will cost a little more and may be more complex to set up, configure, and maintain. But once you are past the initial complexity, the infrastructure architecture provides a more robust solution. A dedicated PC is not required to maintain a wireless home network that allows all workstations access to the Internet.

Infrastructure Mode Wireless Home LAN Using a Hardware-Based NAT

The Buffalo Technologies WLAR-L11-L access point includes a built-in *Dynamic Host Configuration Protocol* (DHCP) server to assign *Internet Protocol* (IP) addresses to the home LAN clients and a NAT that enables one true IP address to be shared among multiple clients on the wired and wireless home network. Windows ICS also provides DHCP and NAT (see Figure 6-3).

In addition to DHCP and NAT, the WLAR-L11-L includes a four-port Ethernet 10/100 built-in hub that enables you to connect home LAN clients via wired as well as wired connections. Using the wired hub of the WLA-L11, more access points can be added to increase the coverage area. In Figure 6-3, a WLA-L11 wireless hub access point is connected to the WLAR-L11-L router access point to illustrate this

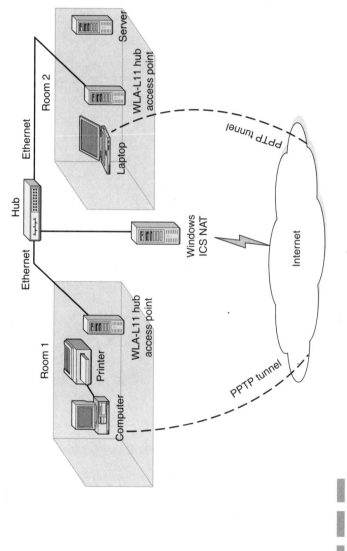

Figure 6-2 ICS-based infrastructure home LAN topology

193

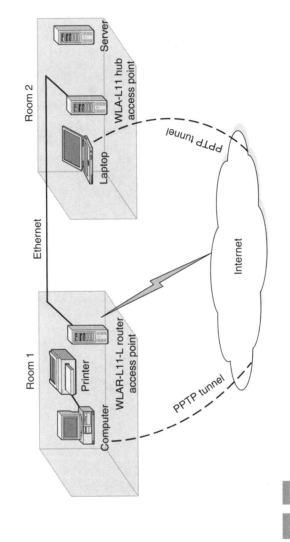

Figure 6-3 Infrastructure mode using the Buffalo Technologies access points

Wireless Home Network Configurations

topology. If you plan to use multiple access points, you can buy one WLAR-L11-L router access point and up to four WLA-L11 hub access points. The hub access point is the cheapest of the two. With this topology, you do not need to use a Windows ICS machine or an extra 10/100 Ethernet hub.

Choosing Between Ad Hoc and Infrastructure Networks

Choose ad hoc if you want

- Lower cost.
- Less complexity.
- You don't have a large house where the WLAN needs to reach more than 100 feet from the location of the ICS system in your home.

Choose infrastructure if you want

- A large coverage area for your WLAN. Typically, this method would be used when a single access point would not provide complete coverage. Given a mostly wooden structure, you can estimate that *radio frequency* (RF) coverage would be adequate up to a radius of 100 feet away from the access point.
- Greater flexibility and control over configuration. The wireless access point includes a proprietary *operating system* (OS) on board that controls every configurable component.

Ad Hoc Network Using ICS

The prerequisites for setting up the ad hoc wireless home LAN are as follows:

- A WLAN card for each of the computers on your home LAN (*Personal Computer Memory Card International Association* [PCMCIA], *Peripheral Component Interface* [PCI], and *Industry Standard Architecture* [ISA] bus versions of cards are available).

- An Internet connection via an analog modem, *Digital Subscriber Line* (DSL), or broadband access. The Internet provider will provide at least one Internet IP address whether it is static or dynamic.
- A Windows ICS machine that could be a PC running Windows 2000 or Windows XP. These OSs include ICS, which allows all the computers on your home LAN to share a single, true IP address. This machine requires both a WLAN card and a second interface that will connect to the Internet. The second interface may be a modem, ISDN, or Ethernet connection to *Asymmetric Digital Subscriber Line* (ADSL) or other broadband Internet access.
- Be sure to have all the adapter software you need before you begin.

Steps to Set Up the NAT Internet Connection and Ad Hoc Wireless Home LAN

Follow the steps below to set up your ad hoc network.

1. Install a WLAN card in each of the PC systems that will operate on your home LAN.
2. Follow all instructions for card installation. If one of these systems is a mobile computer used on the corporate WLAN already, it will not require any reinstallation, software, or hardware. Make sure to check all PCs to ensure that the WLAN setup includes the latest drivers, adapter firmware, and client utilities version available.
3. Go to the Command menu and select Edit Properties. Go to the Home Networking tab, which is shown in Figure 6-4.
4. Enter the following information:
   - **Computer name** Type in the home computer name, which is different for every computer on your home network.
   - *Service set identifier* **(SSID)** This is the SSID name for your home WLAN. It is case sensitive. For security purposes, *this name must be typed in exactly the same on all home systems for successful operation.*

Wireless Home Network Configurations

Figure 6-4
ACU Home Networking tab

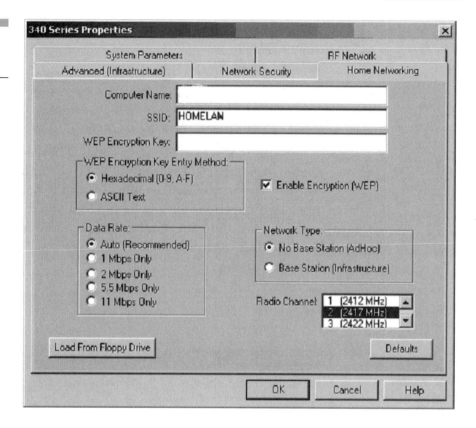

- **WEP encryption key** This is the home *Wired Equivalent Privacy* (WEP) key. It is a 40-bit key using open authentication that is used to encrypt all data between your home systems. Also, machines not using the 40-bit key will not be able to share data. When you first open this, you will see a randomly generated set of numbers. You must type in 10 characters made up of A through F, a through f, and 0 through 9. This will form a 40-bit hex key. *This key must be typed in exactly the same on all home systems for successful operation.*

- **WEP encryption key entry method** You can choose either hexadecimal or ASCII text. Your choice will dictate how many and what type of characters you can use. For instance, if you select Hexadecimal, you will enter 10 characters, which have to be 0 through 9 and/or a through f.

- **Enable data encryption (WEP)** Check this box to enable encryption.
- **Data rate** Leave this on Auto.
- **Network type** Set to No Base Station, which indicates ad hoc mode wireless networking.
- **Radio channel** Leave this on 6.

5. While still in the Edit Properties dialog box, go to the System Parameters tab (see Figure 6-5). You'll see a radio button selector for the current profile. This allows toggling between home and work configurations for any laptop that you may want to use both at work and home. Make sure all home machines including the ICS NAT machine are set to "Use Home Network Configuration."

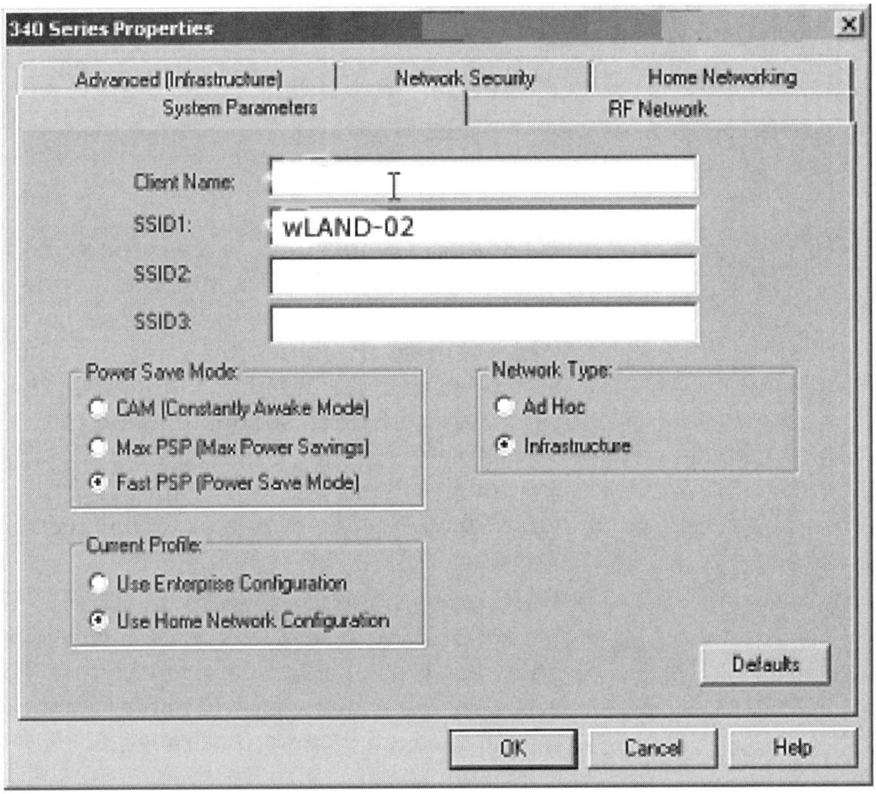

Figure 6-5
ACU System Parameters tab

6. Install the Microsoft TCP/IP on all the computers that will use the Internet on your home network. TCP/IP should be bound to your computer's adapter and set to use DHCP for your IP address and your *domain name system* (DNS) servers. Note that you do not need any protocol except TCP/IP installed on your home LAN-connected computers.
7. Select one of your computers to be your NAT computer. This machine should be running Windows 2000 or the Windows XP OS. All of these OSs include ICS, which offers NAT. In general, you do not need a NAT computer with hardware specifications that are any greater than those indicated for the OS being used. This computer will need two interfaces: (1) the Home LAN wireless adapter interface already installed and (2) a second interface to the Internet. The second Internet interface can be an analog modem, a DSL adapter, or an Ethernet connection to a broadband device.
8. Connect the NAT computer's Internet interface as appropriate. This would mean installation of an analog modem, DSL adapter, or Ethernet card for broadband access to the Internet. If this is a broadband connection, it may require use of a cross-connecting cable. See the broadband service provider's documentation for clarification.
9. Make sure that both interfaces on the NAT computer have the TCP/IP installed. The instance of TCP/IP that is bound to the Internet interface will need to be configured per the instructions of your Internet access provider.
10. As a safeguard against computer hackers, it is recommended that you turn off file sharing on the NAT computer's Internet interface. By connecting to the Internet, you will be exposing your NAT computer's hard drive to the Internet. This puts corporate or personal information on your hard drive at risk. The best way to ensure that your machine is secure is by following directions outlined by the Windows white papers. Windows 2000 and Windows XP are always better choices for security as you can secure all interfaces with separate configurations (you can turn off file sharing on the Internet interface and enable file sharing on the home LAN wireless interface), while file sharing in Windows 9x is a global setting affecting all interfaces.

11. Next, set up ICS. Directions for the setup of ICS are available from www.microsoft.com.
12. After a successful setup, you can connect a home LAN client to the corporate network via *Point-to-Point Tunneling Protocol* (PPTP). Directions for setup are in the "Connecting to the Corporate Network Through Your Home LAN" section.

Test and Troubleshoot the Internet Connection on the Ad Hoc Wireless Home LAN

The last step is to verify that the installation is working properly. Follow the steps below to test the configuration.

1. Open the Command menu on ACU, and select Status on all home PCs, including the ICS NAT System. All PCs should show wireless connectivity via the progress bar indicators at the bottom of the screen (see Figure 6-6).

   Every PC should show Excellent or Good for overall link quality.

   If some PCs show poor connectivity with 0 percent for signal strength, then chances are that they are incorrectly configured. Open the ACU Command menu and check the Home Networking tab to ensure that all configurations on the system are correct. Also, make sure that the test machine is as close as possible to the ICS NAT system to rule out an out-of-range issue.

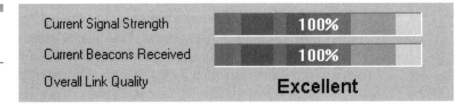

Figure 6-6
ACU status link quality

Wireless Home Network Configurations

2. Now make sure that each of the home clients can receive an IP address from the ICS NAT system. On each of the client systems, open a command window and enter ipconfig/renew at the command prompt. This will force the machine to renew or obtain a new lease on an IP with DNS addressing from the ICS NAT.

- If you cannot get an IP address via DHCP, chances are that there is a problem with the ICS NAT. Determine whether all client PCs are unable to get an IP address from the ICS. If all client PCs cannot get an IP address
  - Check the wireless connection as described in step 1.
  - Reboot the ICS machine and retry ipconfig from the client PCs.
  - Check to make sure that ICS is enabled on the Internet connection. Make sure not to reconfigure your ICS NAT system's Aironet TCP/IP settings. ICS automatically configures these for you.

- If IP configuration succeeds on the client PCs, then you should be able to browse the Internet. Open a web site on one of the PCs. If you cannot open the site successfully
  - Check the cross-connecting cable for *x-Type DSL* (xDSL) or cable modems. In some cases, this problem could be caused by not using a cross-connecting cable. You can identify a cross-connecting cable by holding the two end jacks side by side and examining the colored wires inside the clear plastic connectors. The colored wires of the two jacks should be in the opposite order from left to right. A regular, straight Ethernet cable will show colored wires that go in the same order from left to right. If a cross-connect cable is required for your connection, the broadband service provider will usually supply it along with the modem.
  - Check the ICS NAT system's TCP/IP configuration to make sure it conforms to the *Internet service provider*'s (ISP's) recommended configuration.

Infrastructure Network Setup Using an Access Point Hub and a Windows ICS NAT

The prerequisites for this setup are as follows:

- A Internet connection via analog modem, ISDN, DSL, or broadband access. The Internet provider will provide at least one Internet IP address whether it is static or dynamic.

- A Windows ICS machine that could be a PC running Windows 98 SE, Windows Millennium, Windows 2000, or Windows XP OS. All of these OSs include ICs, which allows all the computers on your home LAN to share a single, true IP address. This machine must have one Ethernet card installed with a wired connection to the airstation and a second connection to the Internet installed (via *Digital Subscriber Line* [DSL], cable, ISDN, modem, and so on). This machine must be running ICS.

- One or more access point(s).

- If using more than one access point in your home, you must have an Ethernet hub (10, 100 or 10/100Base-T).

- A 10Base-T Ethernet adapter installed in the ICS NAT system and connected to the access point, and one standard category 5/RJ-45 network cable to connect the Ethernet adapter to the access point.

- A WLAN card for each of the computers on your home LAN, except the ICS NAT system (PCMCIA, PCI, and ISA bus versions of these cards are available).

- Be sure to have all the adapter software you need before you begin.

NOTE: *CE devices may not be able to seamlessly move from the corporate wireless network to a home. The Windows CE configuration software has no way to manage the differing home and work WEP keys using ACU 4.10. Under ad hoc mode, there are*

Wireless Home Network Configurations

some workarounds though, and Windows CE devices use the Bootstrap Protocol (BOOTP), rather than DHCP, to automatically obtain a working IP address from their network. ICS uses DHCP and does not support BOOTP, so you will need to statically assign an IP address to the card.

Setup of the Buffalo Technologies Airstation WLAR-L11

The following steps are specific to setup of the Buffalo Tech Airstation. The same steps can be followed for use with other access points.

1. Install the Airstation according to the instructions included with the hardware:

   - You can set up your Airstation using a client computer that is connected to it via either a wired Ethernet connection or the wireless connection. If via a wired Ethernet connection, you can set up the Airstation Manager on your ICS computer and connect it to the Ethernet *wide area network* (WAN) port on the Airstation access point. Use the ACU to disable WEP and a blank SSID to make the initial connection to the Airstation using default settings.

   - If you are using only one Airstation, connect the WAN port of the Airstation to the Windows ICS machine.

   - If you have multiple Airstations, connect the WAN links to your Ethernet hub and uplink the hub to the Windows ICS machine.

   - Set the access point WAN uplink to use DHCP for obtaining its IP address.

2. If you are setting up more than one access point in your home, you should try to maximize the overall area covered by all access points and minimize the area in which there is overlapped coverage by the differing access points. Although Buffalo Technology access points support roaming across multiple access points, there is no added feature for detecting and load balancing

between them. The client remains associated with an access point that has a weaker signal, and therefore lower throughput, even after moving close to the nonassociated access point. You should engineer your home LAN such that differing access points have the smallest cell overlap area possible. You can set this up by testing one access point at a time while the others are powered down. Make sure that the approximate area in which one access point's signal strength reaches 0 is where the next access point begins its coverage. This will minimize cell overlap.

3. Configure the Airstation by leaving all defaults except the following:

   - Set up a unique ESS-ID name on the Basic Setup screen and make sure it is the same for all Airstations that you set up in your home (see Figure 6-7).

   - If you are using multiple access points in your home, set Roaming in the Basic Setup screen to Use. Make sure that all Airstations in your home have the same Group Name on the Basic Setup screen.

   - Enable WEP on the Airstation access point.

4. Set up the client adapters on your home computers as directed. Open the Commands menu of the ACU and select Edit Properties. Configure as follows:

   - On the System Parameters tab, select the option for Use Home Network Configuration.

   - On the Home Networking tab, enter the same, case-sensitive SSID in the SSID field that you entered on the Airstation as the ESS-ID.

   - On the Home Networking tab, select the option for Base Station (Infrastructure).

   - On the Home Networking tab, select the checkbox for Enable encryption (WEP), and enter a key.

5. Click OK and the status bar of ACU should indicate an association to the *Medium Access Control* (MAC) address of your

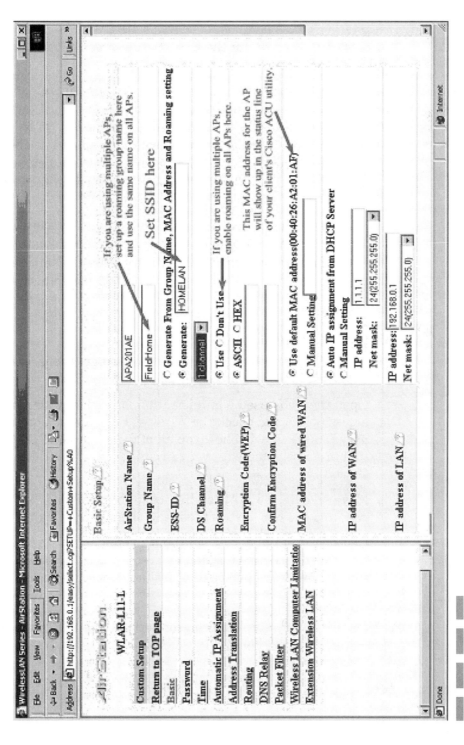

Figure 6-7 Buffalo Basic Setup

Airstation access point. Note that the MAC address of your access point is shown in the Basic Setup screen of the web-based administration client for your Airstation.

6. Open a command like start/run/ and enter ipconfig /all at the command line. You should see an IP address assigned to your Cisco adapter from the range that was specified on your Airstation for automatic IP assignment. You should be able to ping Internet sites such as Yahoo, too. If you do not get a usable IP address, go to the "Troubleshooting Issues with the Airstation WLA-L11/Windows ICS Setup" section.

7. As far as security goes, make sure that you always use an unique SSID on your home network that all clients might share. Note that if your neighbors are close enough (within 100 m) of your house, they may be able to also make an 802.11b connection to your access point. The best way to avoid this is to *only* allow the MAC addresses of your own home LAN cards to associate with your access point. Follow these directions:

   - Turn on all your home WLAN clients and get them connected successfully to your access point.

   - Open the web-based administration for each of your Airstation access points and check the Wireless LAN Client Limitations page. This page will show you all of the client MAC addresses that are associated with your Airstation(s) in the Detected section.

   - Move all detected MAC addresses to the Available list by checking the box next to them and clicking the Change button.

   - Choose the option for Select Limit to enable MAC address filtering and click Set. When you reopen the Wireless LAN Client Limitation page again, it should look like Figure 6-8.

8. Next, set up ICS.

9. After successful setup, you can connect a home LAN client to the corporate network via PPTP.

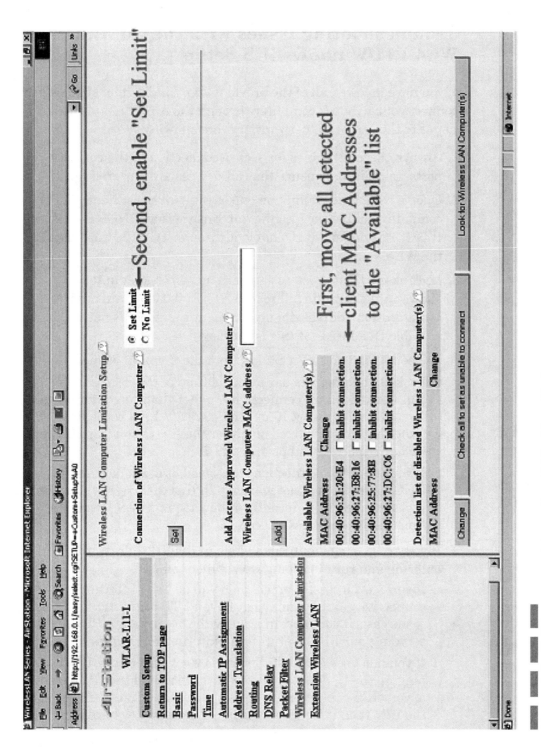

Figure 6-8 Buffalo access point security—MAC address filtering

Troubleshooting Issues with the Airstation WLA-L11/Windows ICS Setup

If you have followed all of the previous steps and cannot successfully connect your wireless computer through the Airstation to the Internet or to the corporate network, try the following:

1. Always check all cables for secure connection to the correct ports. In addition, ensure that all devices are powered up.
2. Open a command prompt on one of your wireless client computers and enter ipconfig /all. Ensure that DHCP is enabled. If DHCP is not enabled, enable it through TCP/IP Properties for the WLAN adapter.
3. Look at the IP address assigned to the wireless adapter as shown in the ipconfig /all results. If it is 0.0.0.0 or an auto-configuration address, then your client computer is not receiving a usable IP address. If this is the case

    - Open ACU and select Status from the Commands menu.
    - Look at the progress bar at the bottom of the Status dialog box and it should show connectivity higher than "poor" for signal strength and quality. If not, you have a WLAN configuration issue that must be fixed on either the Airstation or your client. Go back and retry steps 4, 5, and 7.
    - If your wireless connection looks good, it could be that the DHCP allocation is not working. In this case, retry the setup of ICS, and then retry ipconfig /renew to get a new IP address from the ICS address allocator.

4. If you are getting what looks like a usable IP address, try entering and running "ping www.yahoo.com."
    - If you see successful pings, then you do have Internet connectivity, but you may not have PPTP connectivity. If this is the case, it could be an issue with the corporate PPTP servers, your PPTP setup, or an internetworking issue.
    - If you don't see successful pings to www.yahoo.com, try pinging the IP address of your Airstation and the IP address of your ICS computer. If you can reach the Airstation, but not the ICS, then it is more than likely a cabling or hardware

Wireless Home Network Configurations

issue with the Ethernet ports. If you can get to the ICS, then it is an issue with the DHCP allocator on the ICS machine. Try setting up the ICS again.

If NAT is not the problem, go to the ICS computer and run "ping www.yahoo.com" from a command line. If this fails, you may be experiencing issues with your ISP connectivity and you should contact your ISP.

Infrastructure Network Setup Using an Access Point Router

The prerequisites for this setup are as follows:

- An Internet connection via analog modem, ISDN, or broadband access. The Internet provider will furnish at least one Internet IP address whether it is static or dynamic.
- A Buffalo Airstation WLA-L11-L access point(s).
- If you are using more than one Airstation in your home, you must have an Ethernet hub (10, 100, or 10/100Base-T). For cost effectiveness, additional access points should be WLA-L11 hubs rather than more WLAR-L11-L routers. You only need one router access point to run your network.
- A Cisco 342 WLAN card for each of the computers on your home LAN (PCMCIA, PCI, and ISA bus versions of these cards are available).
- Be sure to have all the WLAN adapter software you need before you begin. You can get all the Cisco/Aironet setup software.

Setup of the Buffalo Technologies Airstation WLAR-L11-L

The following steps are given for the installation of the Buffalo Technologies Airstation access point. Installation of other access points is similar.

1. Install the Airstation according to the instructions included with the hardware:
   - You can set up your Airstation using a client computer that is connected to it via either a wired Ethernet connection or a wireless connection. If via a wired Ethernet connection, connect to one of the ports on the four-port hub of the WLAR-L11-L. If via a wireless connection with a Cisco or Aironet card, use the ACU to disable WEP and a blank SSID to make the initial connection to the Airstation using default settings.
   - If you are using only one Airstation, connect the WAN port of the Airstation to your broadband connection device (DSL or cable modem).
   - If you have opted to use multiple Airstations, you should have a single WLAR-L11-L router and all other access points should be the WLA-L11 hub models. Connect the hub models into the Ethernet four-port hub on the back of the WLAR-L11-L router. If you are using more than four hub access points, you can use a separate, daisy-chained Ethernet hub to host more access points off the router access point.
   - Set the Airstation WAN uplink to use a true IP address from your ISP. This could be either a static IP address or a DHCP-supplied address.
   - If you are setting up more than one airstation in your home, you should try to maximize the overall area covered by all access points and minimize the overlapped coverage by the differing access points. Although Buffalo Technology access points support roaming across multiple access points, there is no added feature for detecting and load balancing between multiple access points. The client remains associated with an access point that has a weaker signal, and therefore lower throughput, even after moving close to the nonassociated access point. You should attempt to engineer your home LAN such that differing access points have the smallest cell overlap area possible. You can set this up by testing one access point at a time while the others are powered down. Make sure that the approximate area in which one access point's signal

Wireless Home Network Configurations

strength reaches 0 is where the next access point begins its coverage. This will minimize cell overlap.

2. Configure the access point by leaving all defaults except the following:

   - Set up a unique ESS-ID name on the Basic Setup screen and make sure it is the same for all Airstations you set up in your home (see Figure 6-9).
   - If you are using multiple access points in your home, set Roaming in the Basic Setup screen to Use. Make sure that all Airstations in your home have the same group name on the Basic Setup screen.
   - If you wish to set up a home WLAN that will enable use of the WLAN adapter that you use in the office, it is recommended that you enable WEP on the access point. The same key must be input on both the Airstation and the client adapter if you use WEP data encryption.

3. Set up your Airstation WLAR-L11-L to provide NAT and DHCP addresses for your Home LAN clients. This function sets up your Airstation to provide the same functions that Windows Internet connection sharing provides.

   - In Custom Setup, click Automatic IP Assignment in the left navigation bar of the web administrator.
   - Select (LAN) Use for Automatic IP Assignment.
   - You may want to specify the addressing range and/or the time limit for Term of Lease.
   - The default gateway and DNS server should be the Airstation's own LAN IP address (by default, it is 192.168.0.1).
   - Click the Set button.
   - Click the Address Translation link in the left navigation bar.
   - Select Use for IP Masquerade Function.
   - Click the Set button.

4. Set up your Cisco 342 client adapters on your home computer. Open up the Commands menu of the ACU and select Edit Properties. Configure as follows:

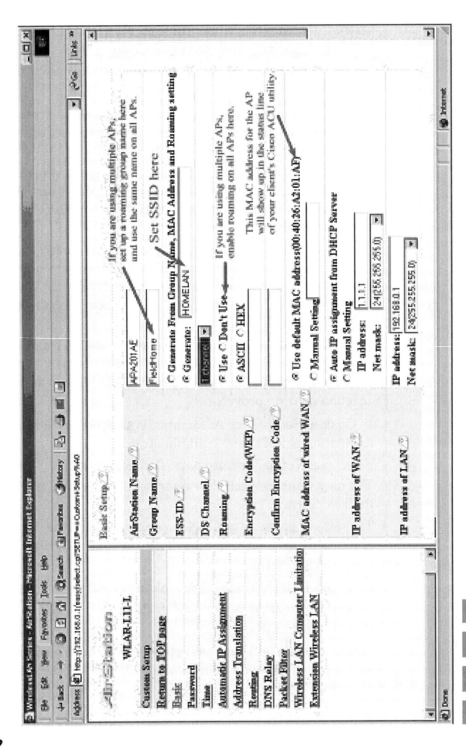

Figure 6-9 Buffalo Basic Setup screen

Wireless Home Network Configurations

- On the System Parameters tab, select the Use Home Network Configuration option.
- On the Home Networking tab, enter the same, case-sensitive SSID in the SSID field that you entered on the Airstation as the ESS-ID.
- On the Home Networking tab, select the Base Station (Infrastructure) option.
- On the Home Networking tab, uncheck (disable) the checkbox for Enable encryption (WEP).

5. Click OK and the status bar of ACU should indicate an association with the MAC address of your Airstation access point. Note that the MAC address of your access point is shown in the Basic Setup screen of the web-based administration client for your Airstation.

6. Open a command like start/run and enter ipconfig /all at the command line. You should see an IP address assigned to your Cisco adapter from the range that was specified on your Airstation for automatic IP assignment. You should be able to ping Internet sites such as Yahoo, too. If you do not get a usable IP address, go to the troubleshooting section at the end of the chapter.

7. Make sure that you always use a unique SSID on your home network that all clients might share. Note that if your neighbors are close enough (within 100 m) of your house, they may be able to also make an 802.11b connection to your access point. The best way to avoid this is to *only* allow the MAC addresses of your own home LAN cards to associate with your access point. Follow these directions:

- Turn on all home WLAN clients and get them connected successfully to your access point.
- Open the web-based administration for each of your Airstation access points and check the Wireless LAN Client Limitations page. This page will show all the client MAC addresses that are associated with your Airstation(s) in the Detected section.
- Move all detected MAC addresses to the Available list by checking the box next to them and clicking the Change button.

- Choose the Select Limit option to enable MAC address filtering and click Set. When you open the Wireless LAN Client Limitation page again, it should look like Figure 6-10.

8. Enable PPTP passthrough on the Airstation WLAR-L11-L from your home network client.

9. After successful setup, you can connect a home LAN client to the corporate network via PPTP.

Troubleshooting Issues with the Airstation WLAR-L11-L Setup

If you have followed all of the previous steps and cannot successfully connect your wireless computer through the Airstation to the Internet or to the corporate network, try the following:

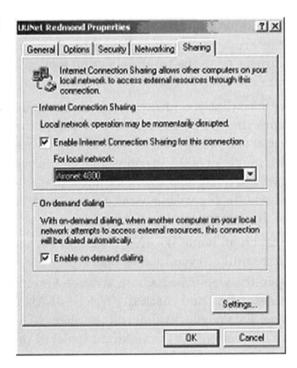

Figure 6-10
Buffalo access point security—MAC address filtering

Wireless Home Network Configurations

1. Always check all cables for secure connection to the correct ports. In addition, ensure that all devices are powered up.
2. Open a command prompt on one of your wireless client computers and enter ipconfig /all. Ensure that DHCP is enabled. If DHCP is not enabled, enable it through TCP/IP properties for the WLAN adapter.
3. Look at the IP address assigned, as shown in the ipconfig /all results. If it is 0.0.0.0 or an auto-configuration address, then your client computer is not receiving a usable IP address. If this is the case:
   - Open ACU and select Status from the Commands menu.
   - Look at the progress bar at the bottom of the page, and it should show connectivity higher than "poor" for signal strength and quality. If not, you have a WLAN configuration issue that must be fixed on either the Airstation or your client. Go back and retry steps 4, 5, and 7.
   - It could be that the DHCP allocation is not working. In this case, retry step 3.
4. If you are getting what looks like a usable IP address, try entering and running ping www.yahoo.com.
   - If you see successful pings, then you do have Internet connectivity, but you may not have PPTP connectivity. If this is the case, go to step 8 for the setup of PPTP passthrough.
   - If you don't see successful pings to www.yahoo.com, try pinging the IP address of your Airstation. If this is successful, then you have an issue with the Airstation NAT or with your Internet access. To fix the NAT, retry step 3.
5. If NAT is not the problem, open the administration web page for the Airstation, click the Diagnosis button, and then the PingTest hyperlink in the left navigation bar. Run a ping to a known Internet IP address, such as 216.32.74.52, which is one of the Yahoo IP addresses. If this fails, you may be experiencing issues with your ISP connectivity and you should contact your ISP.

Windows ICS Setup

This section covers the setup of ICS for Windows 2000 and Windows XP. This section does not cover Windows 2000 or Windows XP *routing and remote access server* (RRAS) NAT setup. It also does not cover the setup of NAT on third-party vendor devices.

These different versions of ICS do have different features, but they all provide NAT. NAT enables the ICS computer to take a single, true IP address and share that address among a group of computers. It does this by assigning a private IP address to each of the computers on the home LAN. When the home LAN computers request Internet content, the ICS computer uses the true IP address to retrieve the content, and then routes it back to the home LAN machine. ICS NAT provides several useful features:

- You only need to purchase one IP address from your ISP, but you can use more than one computer.
- It provides a way to "hide" your home LAN client computers from the Internet. Only the ICS computer's Internet connection interface is exposed to possible security hacking, rather than all of your home LAN computers.
- It allows you to set up an ad hoc home LAN.

In addition to NAT, Windows ICS provides DHCP so that your computers can request and get IP addresses automatically rather than having to hardcode in static IP addresses (see Table 6-1).

ICS Setup for Windows 2000 and Windows XP

Follow these steps to set up Windows 2000 or Windows XP for use as an ICS server.

1. Open Properties for the Internet interface and select the Sharing tab.
2. Check the Enable Internet Connection Sharing for This Connection box.

Table 6-1
Comparison of Windows ICS

Operating System	Ease of Setup	Security/Usability	Support for Multiple Networks
Windows 2000 ICS	Very easy	Good: You can disable file sharing on the Internet interface while keeping it enabled on the home interface.	No, only one home network interface is allowed.
Windows XP ICS	Very easy	Great: You can disable file sharing on the Internet interface while keeping it enabled on the home interface. In addition, you can enable the personal firewall.	Yes, using the new bridging feature, you can enable many different home networks.
Windows 2000/XP RRAS NAT	Complex	Good	Yes, but you must set up a DHCP server with different scopes for each interface.

3. Under For this Network, select the name of the Ethernet card or dial-up that connects to your home network hub. Click OK.

4. If you share a dial-up connection, the On-Demand Dialing section will become visible. Figure 6-11 shows what an ICS configuration would look like in Windows 2000. If the On-Demand Dialing section is not visible it could be due to the following reasons:

   - If you are using an Ethernet connection for xDSL, cable, or other broadband access to the Internet, you will not see the On-Demand Dialing section, as your connection is persistently connected to the Internet.
   - If you only have one network adapter installed on the PC, you will not see For Local Network selection, as shown earlier. This is because ICS automatically uses the only available interface.
   - The Settings button enables you to open or map TCP ports on your NAT for applications that must use those ports. In general, you should never need to use this feature.

Integrating a Wireless Work Network with a Wireless Home Network

Beyond simply roaming within a corporate campus, several other roaming-user scenarios are becoming a reality, with airports and restaurants adding wireless connectivity to the Internet, and wireless networks becoming popular networking solutions for the home.

Now it is becoming more likely that a user could leave the office for a meeting. On the way to this meeting, the user could find himself in a train station, restaurant, or airport with wireless access and need to retrieve files from the office. It would be useful for this user to be authenticated and use this connection to access his corporate network. When the user arrives at his meeting, he may not be given access to the local corporate network he is visiting. It would be fortuitous, however, if the user could be provided with access to the Internet in this foreign environment. This access could then be used

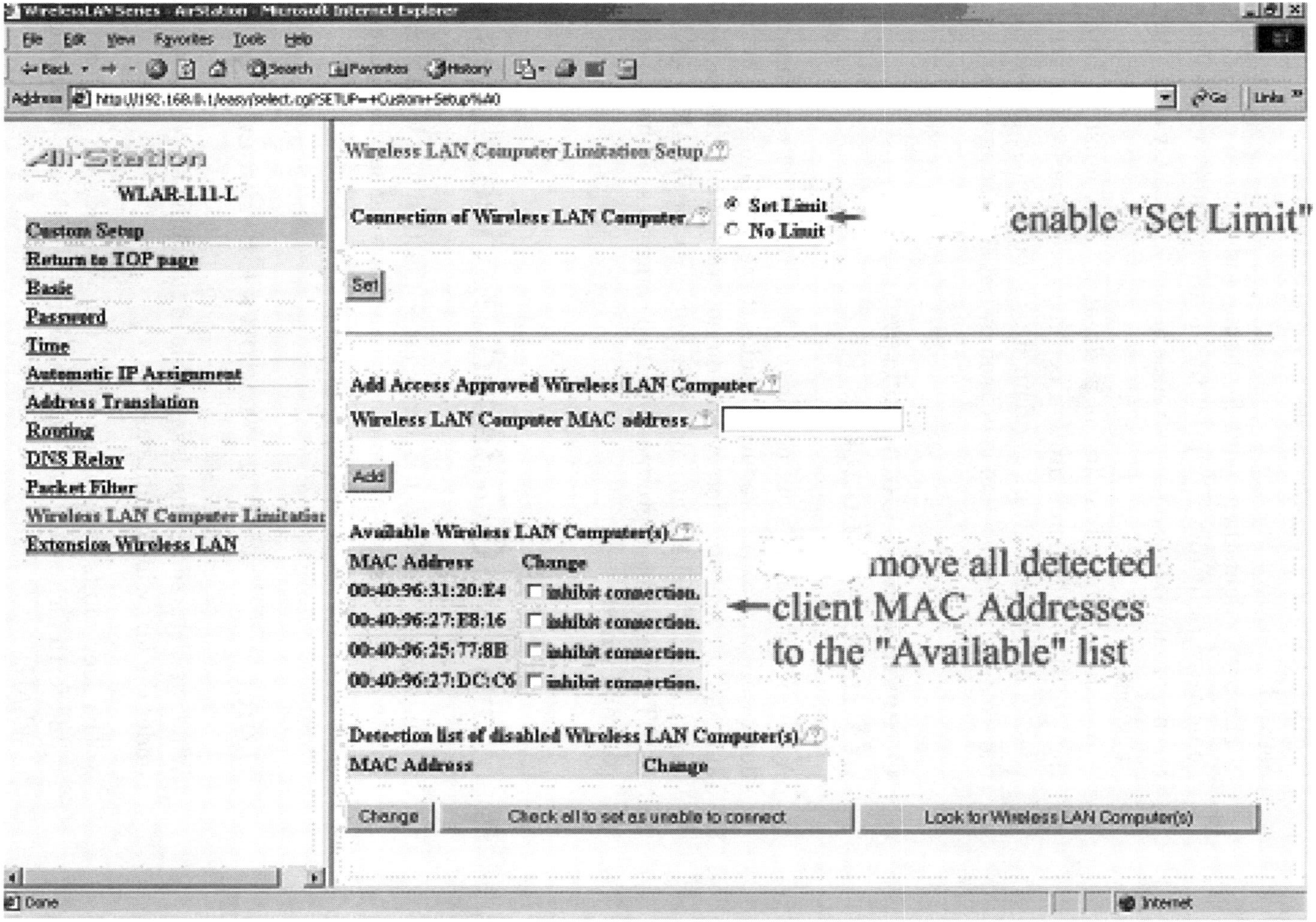

Figure 6-11 CS configuration

to create a *virtual private network* (VPN) connection to his corporate network.

The user might then leave for home and wish to connect to his home network to upload or print files to work that evening. The user has now roamed into a new wireless network, possibly even running in ad hoc mode.

In this example, roaming is a situation that must be carefully thought through. Configuration becomes an issue for the roaming user as different network configurations could cause a challenge if the user's wireless station is not somewhat self configuring.

Overview and Comparison of Wireless Products

A number of products out on the market are 802.11 compliant. You don't want any other products besides those that are 802.11 compliant, as they are proprietary and do not work well with each other (see Table 6-2). There are also different uses for WLANs (see Table 6-3). They have been grouped into two categories: ad hoc or peer-to-peer and wired.

Windows CE-Based Device Notes

Configuring a Windows CE device for connection to a WLAN is similar to other versions of windows. The following sections cover the differences between Windows CE and 2000 or XP.

Using WEP with a Windows CE-Based Device

Windows CE is somewhat less usable in the home LAN environment under the following conditions:

- The user wants to be able to take the Windows CE device back and forth for use on both the corporate WLAN and his home WLAN.

Table 6-2

WLAN Products

Product	Data Rate (MB)	Outside Range (Inside) (Feet)	Interface	Supported OS
Lucent WaveLAN	2 MB	1200 (550)	ISA/PCMCIA	Windows 95/98/NT/2000
BayNetworks BayStack 660	2 MB	2000 (300)	PCMCIA	Novell, Windows 95/98/NT
Samsung MagicLAN	11 MB	1000 (100)	PCI/PCMCIA	Windows 95/98/NT/2000/CE, Linux
Nokia C110/C111	11 MB	1300 (200)	PCI/PCMCIA	Windows 95/98/NT/2000
Cisco Aironet 340	11 MB	400 (100)	ISA/PCI/PCMCIA	Windows 95/98/NT/2000/CE
ZoomAir 4100	11 MB	1000 (300)	PCI/PCMCIA	Windows 95/98/NT/2000
SMC EZConnect	11 MB	1500 (400)	PCI/PCMCIA	Windows 95/98/NT

Table 6-3
WLAN Products

Company	Product	Type	Frequency	Speed	Range
BreezeCom	BreezeNet Pro	Radio FHSS	2.4 GHz	3 Mbps	3,000 feet
Proxim	RangeLAN2	Radio FHSS	2.4 GHz	1.6 Mbps	1,000 feet
Digital	RoamAbout	Radio DSSS and FHSS	915 MHz and 2.4 GHz	2 Mbps	800 feet
WaveAccess	Jaguar	Radio FHSS	2.4 GHz	3.2 Mbps	2,500 feet
IBM	IBM Wireless LAN (withdrawn April '97)	Radio FHSS	2.4 GHz	1.2 Mbps	800 feet
Solectek	AirLAN	Radio DSSS	2.4 GHz	2 Mbps	800 feet
Windata	Freeport	Radio DSSS	2.4 and 5.7 GHz	5.7 Mbps	263 feet
NCR	WaveLAN	Radio DSSS	915 MHz and 2.4 GHz	2 Mbps	800 feet
Aironet	ARLAN	Radio DSSS and FHSS	2.4 GHz	2 Mbps	700 feet
RadioLan	RadioLAN	Microwave	5.8 GHz	10 Mbps	120 feet
Motorola	Altair Plus II	Microwave	18 GHz	5.7 Mbps	250 feet
Photonics		Infrared	N/A	1 Mbps	25' × 25' room
InfraLAN	InfraLAN	Infrared	N/A	16 Mbps	90 feet

Note: Motorola uses frequencies that require licensing from the FCC.

Wireless Home Network Configurations

- The user wants to be able to toggle between the enterprise and home WLAN configurations as he does with the PC configurations recommended earlier.
- The user wants data encryption for security when using his CE device on his home LAN.
- The user wants to access corporate data using his home LAN connection to the Internet.

These issues are mostly the result of the fact that Aironet Cisco currently supports customized home networking for Windows PCs that can run ACU 4.10. There is an offering of Windows CE drivers for StrongArm and *Millions of Instructions Per Second* (MIPS) processors running Windows CE 2.11 HPCPro (Jupiter Class). But the configuration utilities provided with these drivers do not allow for toggling between enterprise and home configurations.

To use a CE device successfully with the recommended home LAN configuration, you would need to enter the home LAN WEP key in the WEP key number one position and overwrite the corporate WEP key.

PC systems running ACU 4.10 or greater can sidestep this problem completely because this version of ACU uses a swap method to put either the corporate WEP key or the home WEP key in the WEP key number one position. Unfortunately, CE cannot do this.

There are several alternative workarounds that you may want to consider though:

- If you only use your CE device at home, then enter the home WEP key in the WEP key number one position. But you will not be able to use this card at work.
- The WEP key is stored in firmware on the physical card. Therefore, you could keep two cards, one for home use with the home WEP key in position number one, and another card with the corporate WEP key in position number one. Note that the software configuration (SSID, encryption type, and so on) would require reconfiguration whenever changing environments.
- You could configure your home LAN to *not* use WEP encryption at all. Then when coming home, you would need to reconfigure the software to turn off encryption altogether on your CE device.

Using DHCP with a Windows CE-Based Device Windows CE devices do not work with all ICS or NAT devices to receive an IP address successfully. This is because they require BOOTP, rather than DHCP, to get an IP address. BOOTP is a legacy IP address allocation protocol that enables the setup of an IP address at the time the device is booted. All of the Windows ICS versions only support DHCP and do not support BOOTP. When you use a CE device on your home LAN with ICS, you can use the following workaround:

1. Select a higher IP address within the subnet of your ICS machine's DHCP addresses. For instance, if you are using the WinMe ICS, you would choose an address between 192.168.0.2 and 192.168.0.254. The ICS DHCP allocator tends to allocate lower addresses, so you can select 192.168.0.254 for your CE device.
2. Open the network applet in the CE device's control panel and enter the IP address you chose and the subnet that your ICS is using.

Connecting to the Corporate Network Through the Home LAN

Once your home LAN is successfully connecting clients through the ICS NAT to the Internet, the clients can now make a connection to the corporate network using VPNs. The connection cannot, however, be made from the ICS NAT computer to a corporate network using a VPN. If the ICS computer is used as a VPN client, it will destroy the network connection of all computers and none will be able to connect. Instead, corporate VPN connections should always be made from one of the home LAN clients (see Figure 6-12).

A VPN will support PPTP connections and other VPN technologies, such as the *Layer Two Tunneling Protocol* (L2TP) and *IP Security* (IPSec), depending on what is deployed on the corporate VPN. Settings for VPN clients should be distributed to the home LAN users.

Wireless Home Network Configurations

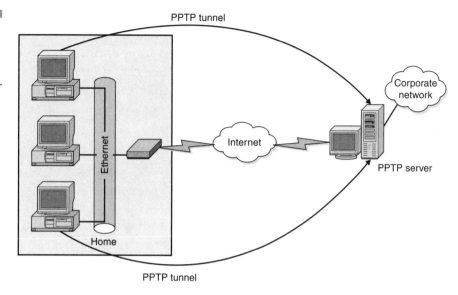

Figure 6-12
Home LAN connectivity to the corporate network

Building the Wireless VPN

VPNs were initially developed to give remote desktop and mobile users, such as telecommuters and on-the-road salespeople, the same sort of secure access to private company networks as people working at company headquarters. Other implementations of VPNs include business partners and employees at branch offices.

VPNs typically run over the public Internet, adding privacy and security through the use of technologies such as tunneling, encryption, authentication, and firewalls.

Summary

A wireless home LAN that easily integrates into the corporate network will make connecting to both networks a breeze. The requirements for a wireless home LAN are slightly different than those of a corporate network. Thought into the standards for home users will go a long way toward providing an easy-to-manage solution. The Buffalo Technologies Airstation and Cisco Aironet NICs were used for testing. The architecture can support any access point and network card combination that provides the needed functionality of DHCP, and NAT.

CHAPTER 7

Wrap Up

Presently, *wireless local area networks* (WLANs) frequently augment rather than replace wired LAN networks, often providing the final few meters of connectivity between a wired network and the mobile user. Some of the many applications and possibilities made possible through the power and flexibility of WLANs are as follows:

- The ability to access LAN resources from any physical point on a corporate premises
- The ability to access corporate data from handheld *personal digital assistants* (PDAs), cell phones, and so on
- The ability to extend corporate data to remote locations, such as warehouses and store rooms, enabling employees to update inventory databases on the fly

In conclusion, there are a million uses and advantages of wireless networks and at present we have only begun to scratch the surface of this rapidly growing technology's potential. Hopefully, this book has provided the IT manager, corporate executive, and other decision-makers with a solid foundation of knowledge, enabling them to sift through the often confusing vocabulary and concepts of wireless networks with confidence. At the very least, we hope that this book has brought awareness to the possibilities of wireless technologies, which undoubtedly are here to stay and will soon be ubiquitous.

As is often the case when exposed to new technological possibilities, one immediately wonders where to begin. Often the question is "What is the first step?" The following is a summary of the steps to take when traveling down the road of the wireless enterprise:

1. Before beginning, ensure that an accurate site survey is readily available.
2. Decide specifically on the functionality desired out of the wireless network.
3. Select hardware based on the in-depth needs analysis.
4. Thoroughly test the hardware for not only vendor compatibility, but compatibility with existing systems.
5. Install and configure a test access point.
6. Update the existing site survey to reflect changes.

Wrap Up

7. Install and configure a wireless client with the appropriate hardware (which in most cases will be a *network interface card* [NIC]).
8. Thoroughly test the test access point and test client, running each through all possible scenarios while recording performance metrics.
9. Install and configure remaining access points.
10. Enable additional features such as security, *service set identifiers* (SSIDs), *Wired Equivalent Privacy* (WEP), and so on.
11. Configure and deploy wireless NICs.
12. Conduct final testing before introducing these components in a production environment.

The history of technology improvements in wired LANs can be summed up with the mantra "faster, better, and cheaper." WLAN technology has already started down that road: data rates have increased from 1 to 11 Mbps, interoperability became reality with the introduction of the IEEE 802.11 standard, and prices have dramatically decreased. The improvements seen so far are just the beginning.

ACRONYMS

AP	Access point
BSS	Basic Service Set
CSMA/CA	Carrier Sense Multiple Access with Collision Avoidance
CSMA/CD	Carrier Sense Multiple Access with Collision Detection
CTS	Clear To Send
DHCP	Dynamic Host Configuration protocol
DS	Distribution system
DSS	Distribution system services
DSSS	Direct sequence spread spectrum
EAP	Extensible Application protocol
ESS	Extended Service Set
FCC	Federal Communications Commission
FCS	Frame check sequence
FHSS	Frequency-hopping spread spectrum
GHz	Gigahertz
IBSS	Independent Basic Service Set
IR	Infrared
LAN	Local area network
LEAP	Lightweight Extensible Application protocol
MAC	Medium Access Control layer
MSDU	MAC service data unit
MW	Microwave
NIC	Network interface card
PAN	Personal area network
PHY	Physical layer
RADIUS	Remote Authentication Dial-In User Service

RF	Radio frequency
RTS	Ready To Send
WEP	Wired Equivalent Privacy
WLAN	Wireless local area network

GLOSSARY

802.11 The IEEE standard that specifies carrier sense media access control and physical-layer specifications for 1 and 2 Mbps WLANs.

802.11b The IEEE standard that specifies carrier sense media access control and physical-layer specifications for 5.5 and 11 Mbps WLANs.

802.1X The new standard for WLAN security as defined by the IEEE. An access point that supports 802.1X and its protocol, Extensible Authentication Protocol (EAP), acts as the interface between a wireless client and an authentication server, such as a Remote Authentication Dial-In User Service (RADIUS) server, to which the access point communicates over the wired network.

Access point A device that transports data between a wireless network and a wired network (infrastructure) using radio waves. This device acts as an Ethernet bridge that handles roaming from cell to cell.

ActiveSync A Microsoft program that enables desktop-to-Windows CE device connections in order to transfer files.

Ad hoc network A wireless network composed of stations without access points.

Alphanumeric A set of characters that contains both letters and numbers.

Associated A station that is configured properly to enable it to wirelessly communicate with an access point.

Backbone Another term for bus, the main wire that connects nodes. The term is often used to describe the main network connections composing the Internet.

Bandwidth Specifies the amount of the frequency spectrum that is usable for a data transfer. It identifies the maximum data rate that a signal can attain on the medium without encountering significant power loss.

BPSK A modulation technique used by IEEE 802.11-compliant wireless LANs for transmission at 1 Mbps.

Bridge A device that connects two LANs or two segments of the same LAN. The two LANs being connected can be alike or dissimilar. For example, a bridge can connect an Ethernet with a token ring network. Unlike routers, bridges are protocol independent. They simply forward packets without analyzing and rerouting messages. Consequently, they're faster than routers but also less versatile.

Cabinet file A self-contained file with a .cab extension used for application installation and setup. In a cabinet file, multiple files are compressed into one file. Cabinet files are commonly found on software distribution disks.

Carrier Sense Multiple Access with Collision Detection (CSMA/CD) The technology that Ethernet is based on. Each device that wants to transmit will first sense whether or not a carrier signal is present on the network. When the network has quit, it will begin to transmit. The device listens for a collision while it transmits the frame; if it detects a collision, it aborts the transmission. Each device involved in the collision schedules its frame for retransmission using a random delay.

CeAppMgr Windows CE Application Manager. This is the desktop Windows CE Services component that provides a desktop-to-device application management tool. It is responsible for adding and removing applications on the Windows CE device and for deleting the application files from the desktop computer. CeAppMgr is included with every installation of Windows CE Services.

cellular digital packet data (CDPD) CDPD is a standards-based wireless technology capable of carrying enough data to compete against personal communications services.

Client A radio device that uses the services of an access point to communicate wirelessly with other devices on a LAN.

Complementary code keying (CCK) A modulation technique used by IEEE 802.11-compliant WLANs for transmission at 5.5 and 11 Mbps.

Glossary

Cyclic Redundancy Check (CRC)   A method of checking for errors in a received packet.

Data rates   The range of data transmission rates supported by a device. Data rates are measured in Mbps.

dBi   A ratio of decibels to an isotropic antenna that is commonly used to measure antenna gain. The greater the dBi value, the higher the gain and the more acute the angle of coverage.

Dipole   A type of low-gain (2.2-dBi) antenna consisting of two (often internal) elements.

Direct sequence spread spectrum (DSSS)   This generates a redundant bit pattern for each bit to be transmitted. This bit pattern is called a chip (or chipping code). The longer the chip, the greater the probability that the original data can be recovered (and, of course, the more bandwidth required). Even if one or more bits in the chip are damaged during transmission, statistical techniques embedded in the radio can recover the original data without the need for retransmission. To the receiver, DSSS appears as low-power wideband noise and is rejected (ignored) by most narrowband receivers.

Domain name server   A network server that translates text names to IP addresses.

Domain name system (DNS)   Provides names for computers using alphanumeric characters instead of numbers like IP addresses use. Maintains a database of the host alphanumeric names and their corresponding IP addresses.

Duplicate packets   Packets that were received twice because an acknowledgement got lost and the sender retransmitted the packet.

Dynamic Host Configuration Protocol (DHCP)   A protocol available with many operating systems that automatically issues IP addresses within a specified range to devices on the network. The device retains the assigned address for a specific administrator-defined period.

Ethernet A LAN protocol developed by Xerox Corporation in cooperation with DEC and Intel in 1976. Ethernet uses a bus or star topology and supports data transfer rates of 10 Mbps. The Ethernet specification served as the basis for the IEEE 802.3 standard, which specifies the physical and lower software layers. Ethernet uses the CSMA/CD access method to handle simultaneous demands. It is one of the most widely implemented LAN standards. A newer version of Ethernet, called 100Base-T (or Fast Ethernet), supports data transfer rates of 100 Mbps. And the newest version, Gigabit Ethernet, supports data rates of 1 Gb (1,000 Mb) per second.

Extensible Authentication Protocol (EAP) The protocol for the optional IEEE 802.1X WLAN security feature. An access point that supports 802.1X and EAP acts as the interface between a wireless client and an authentication server, such as a RADIUS server, to which the access point communicates over the wired network.

File server A repository for files so that a LAN can share files, mail, and programs.

Firmware Software that is programmed on a memory chip and kept in a computer's semipermanent memory.

Fragmentation threshold The size at which packets will be fragmented and transmitted a piece at a time instead of all at once. The setting must be within the range of 64 to 2,312 bytes.

Frequency The number of cycles per unit of time, denoted by Hertz (Hz). One Hz equals one cycle per second.

Gateway A device that connects two otherwise incompatible networks together.

Gigahertz (GHz) One billion cycles per second. A unit of measure for frequency.

Handheld Personal Computer (HPC) One of the three defined types of Windows CE devices.

Hexadecimal A set of characters consisting of 10 numbers and 6 letters (0 through 9, A through F, and a through f).

Glossary

IEEE The Institute of Electrical and Electronic Engineers. A professional society serving electrical engineers through its publications, conferences, and standards development activities. The IEEE is responsible for the Ethernet 802.3 and WLAN 802.11 specifications.

IEEE 802.11 The 802.11 committee standard for 1 and 2 Mbps WLANs. The standard has a single MAC layer for the following physical-layer technologies: frequency-hopping spread spectrum (FHSS), DSSS, and Infrared.

IEEE 802.X A set of specifications for LANs from the IEEE. Most wired networks conform to 802.3, the specification for CSMA/CD-based Ethernet networks. The 802.11 committee completed a standard for 1 and 2 Mbps WLANs in 1997 that has a single MAC layer for the following physical-layer technologies: FHSS, DSSS, and Infrared. IEEE 802.11 HR, an 11 Mbps version of the standard, completed in 1999. Most wired networks conform to 802.3, the specification for CSMA/CD-based Ethernet networks, or 802.5, the specification for token ring networks.

Independent network A network that provides (usually temporarily) peer-to-peer connectivity without relying on a complete network infrastructure or multiple access points.

Infrastructure The wired Ethernet network.

Infrastructure device A device that connects client adapters to a wired LAN, such as an access point, bridge, or base station.

Infrastructure network A wireless network centered about an access point. In this environment, the access point not only provides communication with the wired network but also mediates wireless network traffic in the immediate neighborhood Wireless Local Area Network Alliance (WLANA).

IP address The IP address of a station.

IP subnet mask The number used to identify the IP subnetwork, indicating whether the IP address can be recognized on the LAN or if it must be reached through a gateway.

Isotropic An antenna that radiates its signal 360 degrees both vertically and horizontally in a perfect sphere.

LEAP   Also known as EAP-Cisco Wireless, LEAP is the 802.1X authentication type that is available on Windows CE devices. Support for LEAP is provided in the client adapter's firmware and the Cisco software that supports it, rather than in the operating system. With LEAP, a username and password are used by the client adapter to perform mutual authentication with the RADIUS server through an access point.

Media Access Control (MAC) address   The MAC address is a unique serial number assigned to a networking device by the manufacturer.

Message Integrity Check (MIC)   This prevents bit-flip attacks on encrypted packets. During a bit-flip attack, an intruder intercepts an encrypted message, alters it slightly, and retransmits it; the receiver accepts the retransmitted message as legitimate. The client adapter's driver and firmware must support MIC functionality, and MIC must be enabled on the access point.

Microcell   A bounded physical space in which a number of wireless devices can communicate. Because it is possible to have overlapping cells as well as isolated cells, the boundaries of the cell are established by some rule or convention.

Modulation   Any of several techniques for combining user information with a transmitter's carrier signal.

Multicast packets   Packets transmitted to multiple stations.

Multipath   The signal variation caused when radio signals take multiple paths from transmitter to receiver. In other words, it consists of the echoes created as a radio signal bounces off of physical objects.

Overrun packets   Packets that were discarded because the access point had a temporary overload of packets to handle.

Packet   A basic message unit for communication across a network. A packet usually includes routing information, data, and sometimes error detection information.

Packet-sized Personal Computer (PPC)   One of the three defined types of Windows CE devices.

Glossary

Personal Computer Systems (PCS)   PCS is a new lower-power, high-frequency competitive technology to cellular. PCS operates at the 1.5 to 1.8 Ghz range.

Quadruple Phase Shift Keying   A modulation technique used by IEEE 802.11-compliant WLANs for transmission at 2 Mbps.

Radio channel   The frequency at which a radio operates.

Radio frequency (RF)   A generic term for radio-based technology. The international unit for measuring frequency is Hertz (Hz), which is equivalent to the older unit of cycles per second. One Megahertz (MHz) is one million Hertz. One Gigahertz (GHz) is one billion Hertz. For reference, the standard U.S. electrical power frequency is 60 Hz, the AM broadcast RF band is 0.55 to 1.6 MHz, the FM broadcast RF band is 88 to 108 MHz, and microwave ovens typically operate at 2.45 GHz.

Radio frequency interference (RFI)   Disruption of the signal by radio waves at the same frequency as the desired signal.

Radio spectrum   This consists of radio waves of different frequencies (for example, 900 MHz). All radio spectra are regulated, with some licensed and others unlicensed.

Range   A linear measure of the distance that a transmitter can send a signal.

Receiver sensitivity   A measurement of the weakest signal a receiver can receive and still correctly translate it into data.

Repeater   In a data network, a repeater can relay messages between subnetworks that use different protocols or cable types. Hubs can operate as repeaters by relaying messages to all connected computers. A repeater cannot do the intelligent routing performed by bridges and routers.

Request to send (RTS) threshold   The packet size at which an access point will issue a RTS before sending the packet.

Roaming   The movement of a wireless node between two microcells. Roaming usually occurs in infrastructure networks built around multiple access points. Roaming enables users to move through a facility while maintaining an unbroken connection to the LAN.

RP-TNC   A connector type unique to Cisco Aironet radios and antennas. Part 15.203 of the FCC rules covering spread spectrum devices limits the types of antennas that may be used with transmission equipment. In compliance with this rule, Cisco Aironet, like all other WLAN providers, equips its radios and antennas with a unique connector to prevent attachment of non-approved antennas to radios.

Service Set Identifier (SSID)   A unique identifier that stations must use to be able to communicate with an access point. The SSID can be any alphanumeric entry up to a maximum of 32 characters.

Spread spectrum   Most wireless LAN systems use spread spectrum technology, a wideband RF technique developed by the military for use in reliable, secure, and mission-critical communications systems. Spread spectrum is designed to trade off bandwidth efficiency for reliability, integrity, and security. More bandwidth is consumed than in the case of narrowband transmission, but the tradeoff produces a signal that is, in effect, louder and thus easier to detect, provided that the receiver knows the parameters of the spread spectrum signal being broadcast. If a receiver is not tuned to the right frequency, a spread spectrum signal looks like background noise. There are two types of spread spectrum radio: frequency hopping and direct sequence.

Transmit power   The power level of radio transmission.

Unicast packets   Packets transmitted in point-to-point communication.

WaveLan®   From Wireless Products, this is a complete, integrated system of hardware and software that can extend wireless connectivity to an existing LAN, help you create an instant stand-alone network, and even link the LAN of several buildings. It is a SYSTIMAX® SCS offer.

WavePoint   Lucent technologies access point.

Glossary

WCELoad A Windows CE tool that unpacks a Windows CE .cab file to install an application on a device. It is included with most Windows CE devices.

Wired Equivalent Privacy (WEP) An optional security mechanism defined within the 802.11 standard designed to protect your data as it is transmitted through your wireless network by encrypting it through the use of encryption keys.

Wireless A term that refers to a broad range of technologies that provides mobile communications for the home or office as well as in-building wireless for extended mobility around the work area, campus, or business complex. It is also used to mean cellular for in- or out-of-building mobility services.

Wireless LAN (WLAN) A WLAN is a flexible data communications system implemented as an extension to or as an alternative for a wired LAN. Using RF technology, WLANs transmit and receive data over the air, minimizing the need for wired connections. Thus, WLANs combine data connectivity with user mobility.

Wireless network interface card (NIC) This uses DSSS physical-layer and CSMA/CA medium access control. WaveLAN wireless NIC cards are available in WaveLAN/ISA and WaveLAN/PCMCIA versions.

Wireless node A user computer with a wireless NIC (adapter).

Workstation A computing device with an installed client adapter.

BIBLIOGRAPHY

Balakrishna, Saraswati. *Network Topologies in Wireless LANs.* IFSM 652, December 20, 1995.

Garg, Vijay and Wilkes, Joseph. *Wireless and Personal Communications Systems.* New Jersey: Prentice Hall, 1996.

Gibson, Jerry. *The Communications Handbook.* IEEE Press, CRC Press, 1997.

IEEE P802.11. "Draft Standard for Wireless LAN Medium Access Control and Physical Layer Specification." May 9, 1997.

Muller, Nathan. *Wireless Data Networking.* Norwood, MA: Artech House, 1995.

Pahlavan, Kaveh. "Trends in local wireless data networks," IEEE Vehicular Technology Conference, v1, 1996, pp. 21-25.

Rappaport, Theodore. *Wireless Communications.* New Jersey: Prentice Hall, 1996.

Santamaria, Asuncion. *Wireless LAN Systems.* Norwood, MA: Artech House, 1994.

Simon, Marvin. *Spread Spectrum Communications Handbook.* New York, NY: McGraw Hill Professional Publishing, 1996.

INDEX

Symbols

1000Base-T, 9
100Base-FX, 9
100Base-T, 9–11
10Base-F, 10
10Base-T, 8, 11
10Base2, 7, 10
10Base5, 7, 10
10Broad36, 11
1Base5, 10
3G (third generation), 18, 39–43
4-way handshake, 56
802.11 standard, 25, 29–30, 46, 60–62
802.11a standard, 32
802.11b standard, 31, 74–75
802.11X standard, 124–125
802.1X standard, 77
802.3 standard, 3–4, 7, 14

A

Abramson, Norman, 5, 28
access control, WEP, 75
access point configuration, 98–100
   Ethernet, 107–109
   filter setup, 102–103
   guidelines, 94
   radio, 103–107
   routing, 109–110
   security, 110–115
   WEP, 115–118
access point installation, 86–87
   DHCP servers, 88
   IP addresses, 88–89
   power up, 87
   setup utility, 89
   site surveys, 97
   SSIDs, 89
access point monitoring, 90
access point utilities, 97–98
access pointers, TNC connectors, 91
access points
   antennas, 90
   authentication, 159
   client adapters, 142, 158
   coverage options, 94
   Ethernet ports, 91
   forwarding information, 127
   microcells and roaming, 69
   number of clients, 80
   serial ports, 91
   SOHOs, 63
   troubleshooting, 118–119
   WLANs, 48
ACU, user-level diagnostics, 150
ad hoc home WLANs
   NAT, 196–199
   testing/troubleshooting, 200–201
   via ICS, 195
ad hoc networks, 58
ad hoc WLANs, 61, 165
adapter utility, client adapters, 138
adapters, client. *See* client adapters.
advanced ad hoc parameters, client adapters, 143
advanced configuration, client adapters, 178
advanced infrastructure parameters, client adapters, 141–142
AirPort, 31

Airstation, Buffalo Technologies
  infrastructure setup, 209
  setup, 203–206, 210–213
  troubleshooting, 208, 214–215
Aloha protocol, 28
AlohaNet, 6, 27
AMPS (Advanced Mobile Phone System), 38
antenna modes, client adapters, 141
antenna type, site surveys, 67
antennas, access points, 90
Apple Computer AirPort, 31
association, 59
ATM (Asynchronous Transfer Mode), 16
authentication, 76
  802.11X, 124
  access points, 159
  networks, 112
  RADIUS, 77, 124
  shared secret keys, 60

B

baseband networks, Ethernet, 10
Bluetooth, 33
bridges, wireless, 34
broadband networks, 10
BSS (Basic Service Set), WLANs, 57–62
Buffalo Technologies
  Airstation
    infrastructure setup, 209
    setup, 203–206, 210–213
    troubleshooting, 208, 214–215
  WLA-L11 bridge/hub access point, 191
    infrastructure setup, 209
    setup, 203–206, 210–213
    troubleshooting, 208, 214–215

building-to-building WWANs, 34–35
bus topology, Ethernet, 11

C

cabling, LAN issues, 24
care of addresses, 72
CATV (cable television), 8
CDMA (Code Division Multiple Access), 30, 38–39
CDPD (Cellular Digital Packet Data), 42
central units, wired LANs, 92
channels, client adapters, 140
chip rate, CDMA, 38
chips, DSSS, 51
ciphertext, 60
client adapters, 128–129
  access points, 158–159
  ad hoc WLANs, 131, 165
  client utilities, 131, 165
  configuration, 139–143, 159–161, 178
  CSU (Client Statistics Utility), 180
  current status, 150, 162, 179
  disabling static WEP keys, 150
  drivers, 130, 153–155, 165
  error messages, 186
  firmware upgrades, 152
  infrastructure LANs, 131, 166
  inserting, 152
  install, 132, 167, 170–172
    adapter utility, 138
    client utilities, 169
    drivers, 134–137, 168–169
    unpacking, 132
    verifying, 139, 172
  LEDs, 130
  link quality, 180

Link Status Meter, 151
MAC CRC errors, 182
network configuration, 165
overwriting static WEP keys, 149
radio antennas, 130
radio components, 129
radio firmware, 130, 165
recognition problems, 155
removing, 152
resolving resource conflicts, 156–158
RF link tests, 152
security, 145–148, 160–161
software, 164, 185
troubleshooting, 155, 185
unpacking, 167
up time, 182
user-level diagnostics, 150
viewing statistics, 150, 180
client utilities, 165, 169, 178
clients, access points, 80
coaxial cable, Ethernet, 7
COFDM (coded orthogonal frequency division multiplexing), 32
configuration server protocol, 99
configurations for enterprise, 64
configuring access points, 87
  basic settings, 98–100
  Ethernet, 107–109
  filter setup, 102–103
  radio, 103–107
  routing, 110
  security, 110–115
  WEP, 115–118
  Windows, 162–164
configuring client adapters, 139, 178
  advanced parameters, 141–143
  network security parameters, 143

RF network parameters, 139–141
Windows XP, 159, 161
corporate-home WLAN connections, 224
coverage, 64–68, 94
CSMA/CA (Carrier Sense Multiple Access with Collision Avoidance), 55–56
CSMA/CD (Carrier Sense Multiple Access with Collision Detection), 3, 55–56
CSU (Client Statistics Utility), 180
CTS (Clear To Send), CSMA/CA, 56
current status, client adapters, 150, 162, 179

D

data rates
  3G, 42
  client adapters, 139
  DSSS, 31
  site surveys, 67
data retries, client adapters, 141
deauthentication, 60
default gateways, access points, 100
default IP addresses, access points, 99
default IP subnet masks, access points, 99
DHCP (Dynamic Host Configuration Protocol), 88, 177, 192
diagnostic utilities, 178
diagnostics, client adapters, 150
disabling static WEP keys, 150, 176
disassociation, 59
discovery, SSIDs, 75
distribution, 59
DIX standard, 14
drivers, client adapters, 165
  install, 134–137, 168–169
  upgrades, 153–155

DS (distribution system), 58
DSS (Distribution System Services), 59
DSSS (Direct Sequence Spread Spectrum), 29–31, 48, 51–52
dwell time, FHSS, 50
dynamic EAP keys, 160–161

E

EAP (Extensible Authentication Protocol), 77, 110, 146, 160–161
encapsulation, mobile IP, 72
enterprise configurations, 64
error messages, client adapters, 186
ESS (Extended Service Set), 59, 62
ESS-transition, 59
Ethernet, 3–6, 11, 46
    access points, 107–109
    baseband networks, 10
    Fast Ethernet, 9
    full-duplex, 14
    Gigabit Ethernet, 9
    NICs, 17
    Novell frames, 7
    ports, access points, 91
    segments, length, 8
    Thicknet, 7
    topologies, 11–14
evolution of Ethernet, 46

F

Fast Ethernet, 9
FDDI, 16
FHSS (Frequency Hopping Spread Spectrum), 29, 48–50
filter setup, access points, 102–103
firmware, client adapters, upgrades, 152
FOIRL (Fiber Optic Inter-Repeater Link), 11
foreign agents, mobile IP, 71
fragment thresholds, client adapters, 141
full-duplex Ethernet, 14

G–H

Gigabit Ethernet, 9–11
GSM (Global System for Mobile Communications), 38
hardware, access points, 104, 108
hardware vendors, WLANs, 63
HDML (Handheld Device Markup Language), 37
heavy overlap coverage, 66, 95
help. *See* online help.
hidden node problem, CSMA/CD, 56
home agents, mobile IP, 71
home WLANs, 190–192, 218, 224
HPC devices, online help, 186
HTML (Hypertext Markup Language), 37

I

IAAP (Inter-Access Point Protocol), 127
IAS (Internet Authentication Server), 125
IBSS (Independent Basic Service Set), 58
ICMP Router Discovery, 72
ICS (Internet Connection Sharing), 190–191, 195, 216
IEEE, 3
    802 Executive Committee, 30
    802 standards. *See* 802x standards.
in-building WLANs, 36

Index

infrastructure WLANs, 62–63
   home WLANs, 202
   site requirements, 133
   stations, 57
infrastructure wired LANs, 166
installing access points, 86–89
installing client adapters, 132, 167, 170–172
   adapter utility, 138
   client utilities, 169
   drivers, 134–137, 168–169
   unpacking, 132
   verifying, 139, 172
installing PC cards, Windows devices, 184
installing wireless network, 128–131
integration, 59
interference, WLANs, 21
Internet connections, testing, 200
IP (Internet Protocol), 4
IP addresses
   access points, 88–89
   client adapters, troubleshooting, 185
   home WLANs, 192
IR (infrared), WLAN connections, 2, 52–53
IrDA (Infrared Data Association), 33, 53
ISA client adapters, 36, 48, 129
ISM bands, 19, 30, 47

K–L

key options, WEP, 116–117
LANs, cabling issues, 24
LEAP (Lightweight Extensible Application protocol), 117
link quality, client adapters, 180
Link Status Meter, client adapters, 151
LM cards, client adapters, 142
Lucent WaveLAN, 31

M

MAC address filtering, access points, 102
MAC CRC errors, client adapters, 182
MAC protocols, CSMA/CD, 55
management interface, access point utilities, 98
Metcalfe, Robert, 5
MIC (Message Integrity Check) broadcast key rotation, 111
microcells and roaming, 68
microwave systems, 47, 54
mini-PCI card client adapters, 129, 142
minimal overlap coverage, 64, 94
mobile IP, 70–72
mobile stations, WLANs, 57
mobility, WLANs, 18
monitoring access points, 90
MSDU Delivery, 59
multiple overlapping coverage, 65, 94

N

narrowband, 49
NAT (Network Address Translation)
   ad hoc home WLANs, 196–199
   home WLANs, 192
Negroponte Switch Theory, 55
networks
   ad hoc, 58
   ATM, 16
   authentication types, 112
   bus topology, 11
   configuration, client adapters, 165
   FDDI, 16
   full-duplex, 14
   install, 128–131

integration, 79
security parameters, client adapters, 143
star topologies, 12
switched topology, 14
token ring, 15
NICs (network interface cards), 17
NMT (Nordic Mobile Telephone), 42
no-transition, 59
Novell Ethernet frames, 7

O–P

OIDs (object identifiers), 127
online help, Windows CE devices, 186
open authentication, client adapters, 147
Open System Authentication, 60
OSI model, 3
overwriting static WEP keys, 149, 176
packet-switched data formats, 42
PANs (personal area networks), 33
PARC (Palo Alto Research Center), 4–6
PARs (Project Authorization Requests), 30
PC cards, 129, 141, 184
PCI client adapters, 36, 48, 129, 142
PCMCIA cards, 129
PCS (Personal Communication System), 38
PDAs, wireless connections, 3
PDCP (Personal Digital Cellular Packet), 42
physical environment, site surveys, 67
physical media layer, 802.11 standard, 46
physical-layer extensions, 802.11, 30
plaintext, 60
PnP (Plug-and-Play), 128
Pocket PC devices, connections, 176–177
point-to-point IR systems, 53
portable stations, WLANs, 57
portals, WLANs, 58

ports, Ethernet, access points, 107–108
PPC devices, online help, 186
privacy, WEP, 75
protocols
　Aloha, 28
　filtering, access point configuration, 102
　WAP, 37–38

Q–R

queuing theory, 6
radio
　access point configuration, 103–107
　carriers, 48
　firmware, client adapters, 165
　SSIDs, 100
RADIUS, 77, 111, 123
　802.11X, 124–125
　authentication, 77, 124
range, 68
reassociation, 59
receive statistics, 181
removal
　client adapters, 152
　PC cards, Windows CE devices, 184
repeater units, wired LANs, 92
resource conflicts, client adapters, 156–158
RF (radio frequency)
　client adapters, 139–141, 152
　systems, spread spectrum, 47
　WLAN connections, 2, 25
roaming and microcells, 68
root units, wired LANs, 91–92
RTS (Ready to Send message)
　CSMA/CA, 56
　retry limits, client adapters, 143
RTSD threshold, client adapters, 142

Index

S

seamless roaming, 132, 166
security, 73, 77–79, 173
    access point configuration, 110–115
    authentication, 76
    client adapters, 145–146
    EAP, 160–161
    SSIDs, 74
    static WEP keys, 145–150, 173–176
    WEP, 75, 160
    Windows XP, 123, 128
    WLANs, 110
segments, Ethernet, 8
serial ports, access points, 91
setup, ICS, 216
setup utility, access points, 89
shared address books, 61
Shared Key Authentication, 60, 148
site surveys, 66, 97, 133–134
SNMP options, access points, 101
software, client adapters, 164, 185
SOHOs, access points, 63
spread spectrum, 30, 38, 47–49
spreading code, DSSS, 52
spreading ratio, DSSS, 52
SS (Station Services), 59
SSIDs, 74–75, 89
star topology, Ethernet, 12
static WEP keys, 160
    client adapters, 145–150
    security, 173–176
stations, WLANs, 57
statistics, client adapters, 180
switched topology, Ethernet, 14
system requirements, client adapters, 132–133, 167

T

TACS (Total Access Communications Systems), 42
TCO (total cost of ownership), 37
TDMA (Time Division Multiple Access), 38
testing ad hoc home WLANs, 200–201
Thicknet/Thinnet, 7
throughput, WLANs, 80–82
token ring networks, 15
topologies, Ethernet, 11
transmission media, WLANs, 46
transmit power levels, client adapters, 141
transmit statistics, 183
transparent roaming, 69
troubleshooting
    access points, 118–119
    ad hoc home WLANs, 200–201
    client adapters, 155, 158, 185
    WLAR-L11, 208, 214–215

U–V

UNII (Unlicensed National Information Infrastructure), 32
unpacking client adapters, 132, 167
up time, client adapters, 182
upgrades, client adapters, 152–155
user-level diagnostics, 150, 178
utilities, 97, 138
UTP (unshielded twisted pair), 8
verifying client adapters, 139, 172, 180
viewing statistics, client adapters, 150
VPNs (Virtual Private Networks), 74
    corporate connections, 220
    wireless, 225

W

Walsh codes, CDMA, 39
WAP (Wireless Application Protocol), 37–38
WaveLAN, 31
Web browser interface, access point utilities, 98
WEP, 60, 74–75, 111
   access point configuration, 115–118
   Pocket PC devices, 177
   static keys, 145, 160, 173–176
   using with Windows CE, 220, 223–224
Windows ICS (Internet Connection Sharing), 190–191, 216
Windows 2000
   client adapters, 135, 153, 156–157
   network connections, 125
Windows CE
   configuring access points, 163–164
   devices, 186
      PC card install/removal, 184
      wireless connections, 176–177
   using WEP, 220, 223–224
Windows XP
   client adapters, 154, 157–161
   configuring access points, 162
   drivers, 136–137
   ICS setup, 216
   security, 123, 128
   SSIDs, discovery, 75
   Wireless Configuration Zero service, 127
   wireless network configuration, 126
   WLAN support, 122
wired Ethernet, 7
wired LANs, 91–92
Wireless Configuration Zero service, Windows XP, 127
WLA-L11 bridge/hub access point, 191
   setup, 203–206, 209–213
   troubleshooting, 208, 214–215
WLANs
   802.11 standard, 25
   access points, 48
   ad hoc, 61, 200–201
   adapters, 48
   authentication, 76
   BSS (Basic Service Set), 57
   DS (distribution system), 58
   DSS (Distribution System Services), 59
   ESS (Extended Service Set), 59
   home use, 190–192
   in-building, 36
   infrastructure, 57, 62–63, 202
   portals, 58
   RADIUS, 123
   security, 73, 77–79, 110
   SS (Station Services), 59
   SSIDs, 74
   throughput, 80–82
   transmission media, 46
   WEP, 74–75
   Windows XP support, 122–123
WML (Wireless Markup Language), 37
WMLScript, 37
World mode, client adapters, 140
WPANs (wireless PANs), 33
WWANs (wireless wide area networks), 28, 34–35

X–Z

Xerox, 6
zero configuration, Windows XP, 127

ABOUT THE AUTHOR

Jaidev Bhola formerly was a consultant with Microsoft's Telecom Practice and currently is a principal at a wireless security company. He has over a decade's experience as a Windows administrator, developer, and network engineer. Mr. Bhola also holds CompTIA A+, MCSE 2000, and CCNA certifications.